Coastal Resources Economics and Ecosystem Valuation

Coastal Resources Economics and Ecosystem Valuation

Special Issue Editors

J. Walter Milon
Sergio Alvarez

MDPI • Basel • Beijing • Wuhan • Barcelona • Belgrade

MDPI

Special Issue Editors

J. Walter Milon
University of Central Florida
USA

Sergio Alvarez
University of Central Florida
USA

Editorial Office
MDPI
St. Alban-Anlage 66
4052 Basel, Switzerland

This is a reprint of articles from the Special Issue published online in the open access journal *Water* (ISSN 2073-4441) in 2019 (available at: https://www.mdpi.com/journal/water/special_issues/ Coastal_economics_ecosystem).

For citation purposes, cite each article independently as indicated on the article page online and as indicated below:

LastName, A.A.; LastName, B.B.; LastName, C.C. Article Title. *Journal Name* **Year**, *Article Number, Page Range.*

ISBN 978-3-03928-016-2 (Pbk)
ISBN 978-3-03928-017-9 (PDF)

Cover image courtesy of Grant Leslie.

Contents

About the Special Issue Editors

J. Walter Milon is a Provost's Distinguished Research Professor in the Deoartment of Economics and a founding member of the National Center for Integrated Coastal Research. He has over 40 years of experience in water resource economics, ecosystem valuation, and environmental policy. In addition to his academic research and publications, Dr. Milon has conducted research and consulting for a number of federal agencies including the Environmental Protection Agency, the National Marine Fisheries Service, the National Oceanic and Atmospheric Administration, the National Research Council, and the U.S. Army Corps of Engineers.

Sergio Alvarez is an Assistant Professor at the Rosen College of Hospitality Management and the Sustainable Coastal Systems Cluster at the University of Central Florida. He is an economist researching how natural resources and the environment contribute to human well-being through the provision of ecosystem services such as food, recreation, and protection from natural and man-made hazards.

Editorial

Coastal Resources Economics and Ecosystem Valuation

J. Walter Milon [1],* and Sergio Alvarez [2]

[1]   Department of Economics, University of Central Florida, Orlando, FL 32816, USA
[2]   Rosen College of Hospitality Management, University of Central Florida, Orlando, FL 32816, USA;
      sergio.alvarez@ucf.edu
*   Correspondence: jmilon@ucf.edu

Received: 11 October 2019; Accepted: 16 October 2019; Published: 23 October 2019

Abstract: The papers in this special issue provide new insights into ongoing research to value coastal and marine ecosystem services, and offer meaningful information for policymakers and resource managers about the economic significance of coastal resources for planning, restoration, and damage assessment. Study areas encompass a broad geographic scope from the Gulf of Mexico in the United States, to the Caribbean, the European Union, Australia, and Southeast Asia. The focus of these papers ranges from theoretical perspectives on linkages between ecosystem services and resource management, to the actual integration of valuation information in coastal and marine resource policy decisions, and to the application of economic valuation methods to specific coastal and marine resource management problems. We hope readers will appreciate these new contributions to the growing literature on coastal and marine resource ecosystem services valuation.

Keywords: environmental valuation; coastal ecosystem services valuation; coastal management; ecosystem restoration

---

1. Introduction

Coastal areas around the world are dynamic environments at the interface of terrestrial, marine, and freshwater systems. Nearly 2.4 billion or 40% of the world's people already live in these areas [1]. Coastal zones are increasingly attractive for development and tourism. However, the coastal ecosystems within these zones are vulnerable to a variety of impacts from anthropogenic activities, resulting in excess nutrients, invasive species, extreme weather, sea level rise, and oil spills, among others. These coastal ecosystems include, but are not limited to: estuaries, beaches, wetlands, shores, mangroves, seagrasses and salt marsh, coral reefs, and other essential habitats for marine life.

This special issue focuses on economic valuation of coastal and marine resource ecosystem services. Economic valuation is important because it provides methods and techniques to determine how changes in coastal and marine ecosystem services can be translated into benefits and costs to society. Economic values play an important role in everyday life and provide useful information about human happiness and welfare. Valuation provides a consistent framework to understand human–nature interactions across a broad range of coastal and marine resources, and to evaluate the sustainability of these interactions. The focus on ecosystem services provides new research on this innovative perspective on human–nature interactions that has profoundly changed the academic dialogue on natural systems, but has had limited impact on public dialogue and the policy process.

The practical importance of economic valuation information can hardly be overstated. Coastal and marine resource policy planning and management benefit from complete information on the impact of policy decisions. In addition, proper accounting of the impacts of these policy decisions is necessary for benefit-cost analyses and measurements of economic growth over time.

The papers in this special issue provide new insights into ongoing research to value coastal and marine ecosystem services, and offer meaningful information for policymakers and resource managers about the economic significance of coastal resources for planning, restoration, and damage assessment. Study areas encompass a broad geographic scope from the Gulf of Mexico in the United States, to the Caribbean, the European Union, Australia, and Southeast Asia. The focus of these papers ranges from theoretical perspectives on linkages between ecosystem services and resource management, to the actual integration of valuation information in coastal and marine resource policy decisions, and to the application of economic valuation methods to specific coastal and marine resource management problems. We hope readers will appreciate these new contributions to the growing literature on coastal and marine resource ecosystem services valuation.

2. Summary of the Papers

This special issue contains seven papers. The first paper by Milon and Alvarez [2] provides an overview of the literature on valuation of coastal and marine ecosystem services, and the applications of valuation studies in policy and planning across the world. The authors begin with a description of the ecosystem services concept in the context of coastal and marine ecosystems, and the linkages between these ecosystem services and economic measures of use and nonuse values. The article focuses on prior literature that has attempted to identify economic values for ecosystem services and the use of these studies to estimate global values for coastal and marine ecosystems. This review indicates that there are significant gaps in the existing research, and there are few new efforts to fully integrate the relationships and feedbacks between ecosystems and the services they produce into economic valuation. A review of studies focusing on the application of economic valuation information for coastal and marine resource planning and policy in the United States, the Caribbean, the European Union, and Australia, reveals that valuation information is not widely understood and has had a negligible impact on the policy process. The authors conclude that the application and use of economic valuation information for coastal and marine resource planning and policy is not likely to advance until a more encompassing framework, such as wealth accounting, is adopted to evaluate human–nature interactions, the broad range of services provided by these ecosystems, and the impact of management decisions on sustainability.

Barbier [3] develops a theoretical framework to consider the impact of an open access fishery harvesting regulation on coastal habitat-dependent fishery stocks. The bioeconomic model integrates coastal breeding and nursery habitat availability with quota rules to limit the harvest of near-shore stocks. The model indicates that fixed quota rules fail to capture the changes in economic value due to interdependence between the habitat and fishery stocks, and a flexible management regime is necessary to achieve maximum economic yield. The modeling framework is applied to mangrove-dependent shellfish and demersal fishery species in Thailand. This analysis identifies significant differences in economic welfare when quota rules are not adjusted, and these effects vary across shellfish and demersal species. The article provides an important rationale for coastal and marine resource valuation studies to address the role of identifying and managing habitat to fishery linkages, and the dynamic relationships between habitat development and harvesting regulations.

Alvarez et al. [4] develop a random utility model to simulate changing recreational boating site choices as a result of harmful algae blooms (HABs) in coastal waters. The approach relies on survey data collected in the past to identify boater preferences for particular access ramps throughout a coastal county of the United States. The model controls for site attributes such as presence of artificial reefs, navigation aids, and no-wake zones to protect manatees, as well as protected status and water depth. The model is used to value site closures observed during recent cyanobacterial HABs caused by excess nutrients from human sources. Due to these nutrient loads and in synergy with historic surface water management (e.g., dredging and diking from urbanization and agriculture), the study area is suffering from chronic, semi-annual toxic HABs. Besides contributing to an emerging literature on the negative values of HABs, the article provides timely information on a topic of heightened public and policy-maker attention.

Schuhmann et al. [5] develop a stated behavior approach to measure visitors' willingness to return to Barbados in the future, using scenarios with different water quality, beach width, and coral reef health. While they find that visitors' intentions to return are impacted by changes in the three attributes examined, it is water quality that has the largest impact on intentions to return to Barbados. Further, large portions of respondents who had expressed they would definitely or probably return, later stated that even with small reductions in water quality they would definitely or probably not return. Their findings demonstrate that tourists' return visitation decisions are sensitive to declines in environmental quality, and provide justification for investments or regulations designed to maintain or improve environmental quality as measures to ensure the vitality of tourism in the region.

Maynard et al. [6] use a contingent valuation approach to value changes in coral ecosystems in a protected area in Taiwan that has experienced rapid and drastic deterioration as a result of nutrient pollution, overfishing, extreme weather, and coral bleaching. The approach considers both improvements in coral reef quality through restoration, as well as declines in coral reef quality due to further deterioration of the coral reef ecosystems in the protected area. To ensure that the respondents had an adequate understanding of the scenarios being considered, the approach uses photographs from areas within the protected area showing a progression of low to high coral coverage, and contribution to a coral protection trust is used as the payment vehicle. Their results indicate that respondents' willingness to pay for coral conservation increases with increasing coral coverage, and conversely, the marginal costs of degradation decrease with increasing coral coverage.

Seidel, Dourte, and Diamond [7] evaluate the use of remote sensing data in economic valuation using case studies across the Gulf of Mexico. Although remote sensing data is increasingly used for terrestrial ecosystem studies, applications for coastal and marine ecosystems have been limited. Their analysis is based on workshops with coastal managers and researchers across the Gulf States and valuation studies at National Estuary Program sites along the Atlantic and Gulf coasts of Florida. The workshops identified several barriers to adoption of remote sensing data in the Gulf region including: temporal and spatial gaps in the existing data, uncertainty about the precision of remote sensing data for coastal and marine resources, and the availability of economic valuation data for coastal and marine ecosystem services. The case studies focused on coastal resiliency and habitat restoration for coastal wetland and mangrove ecosystems. The valuation component employed a variety of benefit transfer methods including the use of InVEST, an ecosystem services valuation based software program that utilizes spatial data. Their evaluation indicates that remote sensing data can be successfully integrated into coastal and marine resource valuation studies, however, significant barriers must be overcome.

Conflicts of Interest: The authors declare no conflict of interest.

References

1. United Nations. Factsheet: People and Oceans. In Proceedings of the Ocean Conference, New York, NY, USA, 5–6 June 2017; Available online: https://www.un.org/sustainabledevelopment/wp-content/uploads/2017/05/Ocean-fact-sheet-package.pdf (accessed on 9 October 2019).
2. Milon, J.W.; Alvarez, S. The elusive quest for valuation of coastal and marine ecosystem services. *Water* **2019**, *7*, 1518. [CrossRef]
3. Barbier, E.B. Valuing coastal habitat-fishery linkages under regulated open access. *Water* **2019**, *4*, 847. [CrossRef]
4. Alvarez, S.; Lupi, F.; Solis, D.; Thomas, M. Valuing provision scenarios of coastal ecosystem services: the case of boat ramp closures due to harmful algae blooms in Florida. *Water* **2019**, *6*, 1250. [CrossRef]
5. Schuhumann, P.; Skeete, R.; Waite, R.; Bangwayo-Skeete, P.; Casey, J.; Oxenford, H.A.; Gill, D.A. Coastal and marine quality and tourists' stated intention to return to Barbados. *Water* **2019**, *6*, 1265. [CrossRef]

6.   Maynard, N.; Chateau, P.A.; Ribas-Deulofeu, L.; Liou, J.L. Using internet surveys to estimate visitors' willingness to pay for coral reef conservation in the Kenting National Park, Taiwan. *Water* **2019**, *7*, 1411. [CrossRef]

7.   Seidel, V.; Dourte, D.; Diamond, C. Applying spatial mapping of remotely sensed data to valuation of coastal ecosystem services in the Gulf of Mexico. *Water* **2019**, *6*, 1179. [CrossRef]

water

MDPI

Review

The Elusive Quest for Valuation of Coastal and Marine Ecosystem Services

J. Walter Milon [1,*] and Sergio Alvarez [2]

[1]   Department of Economics, University of Central Florida, Orlando, FL 32816, USA
[2]   Rosen College of Hospitality Management, University of Central Florida, Orlando, FL 32816, USA
*   Correspondence: jmilon@ucf.edu

Received: 3 June 2019; Accepted: 15 July 2019; Published: 22 July 2019

Abstract: Coastal and marine ecosystem (CME) services provide benefits to people through direct goods and services that may be harvested or enjoyed in situ and indirect services that regulate and support biological and geophysical processes now and in the future. In the past two decades, there has been an increase in the number of studies and journal articles designed to measure the economic value of the world's CME services, although there is significantly less published research than for terrestrial ecosystems. This article provides a review of the literature on valuation of CME services along with a discussion of the theoretical and practical challenges that must be overcome to utilize valuation results in CME policy and planning at local, regional, and global scales. The review reveals that significant gaps exist in research and understanding of the broad range of CME services and their economic values. It also raises questions about the validity of aggregating ecosystem services as independent components to determine the value of a biome when there is little understanding of the relationships and feedbacks between ecosystems and the services they produce. Finally, the review indicates that economic valuation of CME services has had a negligible impact on the policy process in four main regions around the world. An alternative direction for CME services research would focus on valuing the world's CME services in a wealth accounting framework.

Keywords: coastal; marine; ecosystem services; economic valuation; wealth accounting; public policy

1. Introduction

Coastal regions around the world have been important centers for population growth and economic development, and this trend is projected to continue throughout the 21st century [1]. Although coastal areas cover only 4% of the Earth's total land area and 11% of the ocean area, they contain more than a third of the Earth's human population and are more than twice as densely populated as inland areas [2,3]. The land–sea interface and the coastal and marine ecosystems (CMEs) within this interface provide a rich array of goods and services ranging from fish and shellfish harvesting in seagrass beds, mangroves, oyster reefs, and coastal bays and estuaries to recreation and tourism on beaches, shores, and coral reefs. Past development, however, has led to significant loss and degradation of these CMEs [4,5]. Costanza et al. [6] estimated that the global supply of CMEs, with the exception of seagrass beds, had exhibited significant losses in both area and total economic values between 1997 and 2011. However, recent research indicates that seagrass beds have also experienced a marked global decline in terms of area covered [7]. Future demographic trends and sea-level rise will intensify the pressures for change in these ecosystems [1].

A number of national and international policies have been implemented to protect and restore CMEs and the ecosystem services they provide [8–12]. Following the Millennium Ecosystem Assessment [2] framework, these ecosystem services provide benefits to people through direct goods and services that may be harvested or enjoyed in situ and indirect services that regulate and support biological and geophysical processes such as nutrient cycling, water purification, and reproductive habitats now and in the future [13]. A number of different classification systems have been developed in recent years to define the linkages between ecosystem services and benefits to people including the National Ecosystem Services Classification System [14], the Common International Classification of Ecosystem Services [15], and the United Nations System of Environmental-Economic Accounting [16]. In addition, the International Organization for Standardization has recently issued ISO 14,008, which specifies a methodological framework for the monetary valuation of environmental and related impacts on human health, the built environment, and ecosystems [17].

A critical component in identifying the contribution of ecosystem services from CMEs is quantification of these contributions as economic values or similar metrics. Economic valuation is increasingly recognized as a necessary component of CME evaluation and policy analysis because it provides a commensurate metric to compare different ecosystem services and tradeoffs between these services and other economic activities that may impact CMEs [8]. In the past two decades, there has been an increase in the number of studies and journal articles designed to measure the economic value of CMEs, although there are significantly fewer published coastal and marine articles than for terrestrial ecosystems [18,19]. Access to the results of these studies is now available through online databases such as the National Ocean Economics Program/Middlebury Institute of International Studies at Monterey, Economics of Ecosystems and Biodiversity, Environmental Valuation Reference Inventory, and the Gulf of Mexico Ecosystem Services Valuation Database. In addition, ecosystem services models such as the Integrated Valuation of Ecosystem Services and Tradeoffs (InVEST) have modules that can be applied for CMEs.

In this article, we provide a review of the literature on valuation of CME services along with a discussion of the theoretical and practical challenges that must be overcome to utilize valuation results in CME policy and planning at large scales. We begin in the next section with an overview of the different coastal and marine ecosystems and the provisioning, cultural, regulating, and supporting services provided by these ecosystems. We also relate these services to specific types of direct, indirect, and nonuse values for each CME. Section 3 reviews the economic research to value CME services in terms of the focus on specific types of CMEs and services. This review indicates that past research has focused heavily on coral reefs and mangroves and the provisioning services they provide. Other CMEs such as seagrass beds and oyster reefs have not received as much attention, and the values associated with regulating and supporting services for CMEs are not well understood. Section 4 considers the role of valuation in the CME policy and planning process in four regions across the world and the impediments to successful integration of CME services valuation. The final section considers future directions for CME services valuation and the difficult challenges associated with a complete accounting for the Earth's CME services.

2. The Scope of Coastal and Marine Ecosystem Services and Economic Valuation

Coastal and marine ecosystems and the processes they support produce flows of services that are highly valuable to society [3,14]. As long as they are not degraded or depleted, these ecosystems will continue to produce services that offer a variety of benefits to people. These benefits can be measured in monetary units using total economic values (TEVs) that recognize the direct, indirect, and nonuse contributions of these ecosystem services to humanity e.g., [20]. However, this monetary valuation may be difficult and may not capture the full range of ecological, socio-cultural, and non-anthropocentric values that can be ascribed to natural capital such as CMEs [21,22]. Nevertheless, one advantage of monetary valuation is that it can potentially be used within a framework that consistently evaluates both conventional economic activity and changes in the stock of natural capital within a unified wealth accounting system e.g., [16,23,24]. Similarly, monetary valuation of ecosystem services can be used to examine the impact of specific policies or management interventions affecting these ecosystems through benefit–cost analysis [25].

2.1. Ecosystem Services in the Economic Valuation Framework

The conceptual work emanating from the Millennium Ecosystem Assessment [2] was among the first efforts to define a comprehensive classification system for the types of services provided by ecosystems. These ecosystem services can be grouped into four categories: provisioning, cultural, regulating, and supporting. Provisioning services provide food and raw materials that can be directly consumed or utilized by humans. Cultural services are nonmaterial benefits that are enjoyed through recreation and aesthetic experiences, spiritual, or artistic appreciation. Regulating services have a less direct (albeit no less important) impact on humans through processes such as water treatment or purification, carbon storage, hydrologic regulation, flood protection, wave attenuation, and erosion prevention. Supporting services are necessary for the production of other ecosystem services and include natural habitats, biological diversity, and climate stability. These categories are not mutually exclusive, and the continuing supply of ecosystem services depends on healthy interaction and maintenance of all stocks of natural capital.

The TEV is the most common valuation framework to capture the range of benefits from ecosystem services, and it provides an overview of the benefits that humans receive from ecosystems and their services as well as the motivations that people may have for wanting to conserve and preserve ecosystems. In other words, TEV is an anthropocentric concept that considers economic value strictly as physical or perceived benefits to humans, a point which has fueled disagreements between economists and conservationists on the issue of intrinsic value [26]. TEV is the sum of benefits that derive from direct use value (UV), indirect use value (IUV), and nonuse value (NUV). These can be further refined into more specific subcategories e.g., [20], but these distinctions are most relevant in the consideration of future supply of ecosystem services. In the context of CMEs, UVs come from the harvesting of fish, wildlife, and raw materials in different CME habitats as well as activities that involve non-consumptive recreational uses such as snorkeling, scuba diving, and nature observation. Regulating services such as storm protection and erosion control provide IUVs. Supporting services such as biodiversity and ecological connectivity that allow species to move across habitats at different stages of their life cycle are not directly or indirectly consumed but provide NUV through their role in maintaining ecosystem functions and the future availability of ecosystem services.

A more complete classification of economic values for ecosystem services from different types of CMEs is presented in Table 1. For each category of ecosystem services, UV and IUV benefits are noted in regular font and NUVs are in italics. The geographic scale at which these benefits are typically received is denoted along the bottom of the table. The most commonly recognized benefits are provisioning and cultural services that are directly used, most often, at the local or regional scale. As discussed in the following section, these are the ecosystem services from CMEs that have been most widely studied for economic valuation. Regulating services such as the storm protection benefits of mangroves, beaches, and dunes also provide UV and IUV and have received some attention in the valuation literature. Other types of regulating services of CMEs such as providing nursery and protective habitat for a range of fish and wildlife species are less understood and more difficult to value. Typically, these services function at a broader regional scale, so identifying the 'extra-local' linkages across ecosystems is necessary for services valuation e.g., [27].

Table 1. Coastal ecosystems and types of economic values for ecosystem services. Direct and indirect use values in regular type, nonuse values in italics.

Coastal Ecosystem	Ecosystem Services		
	Provisioning and Cultural	Regulating	Supporting
Coral Reefs	Recreation and tourism, fish and shellfish harvesting, raw materials, *education* and *aesthetics*	Storm protection, nutrient cycling	*Biological diversity, ecological connectivity, habitat for fish and shellfish, nursery and protective habitat*
Seagrass beds and salt marshes	Fish and shellfish harvesting, raw materials, wildlife harvesting, *education* and *aesthetics*	Storm protection, erosion control, water purification, oxygen cycling, nutrient cycling, carbon storage and sequestration	*Biological diversity, ecological connectivity, nursery and protective habitat for fish, shellfish and wildlife*
Mangroves	Fish and shellfish harvesting, raw materials, *education* and *aesthetics*	Storm protection, nutrient cycling and erosion control, water purification, oxygen cycling, carbon storage and sequestration	*Biological diversity, ecological connectivity, nursery and protective habitat for fish, shellfish and wildlife*
Oyster reefs	Shellfish harvesting, raw materials, *education* and *aesthetics*	Storm protection, erosion control, water purification, nutrient cycling, carbon storage and sequestration	*Biological diversity, ecological connectivity, nursery and protective habitat for fish and shellfish*
Beach, dune and shore	Recreation and tourism, *education* and *aesthetics*	Storm protection, erosion control	*Biological diversity, ecological connectivity, nursery and protective habitat for shellfish and wildlife*
Bays and estuaries	Recreation and tourism, fish and shellfish harvesting, raw materials, wildlife harvesting, *education* and *aesthetics*	Storm protection, erosion control, water purification, oxygen cycling, carbon storage and sequestration, nutrient cycling	*Biological diversity, ecological connectivity, nursery and protective habitat for fish, shellfish and wildlife* (see [4] p. 187)
Local		**Scale**	*Global*

At the broadest, or global scale, supporting services from CMEs produce NUV for processes such as biodiversity and carbon sequestration. These are ongoing processes at the regional and global scale that are necessary for the long-term survival of CMEs. These processes, however, are difficult to characterize, and there is more uncertainty about their functioning. For example, biodiversity improves the capacity of ecosystems to adapt to climate change, but the changes in species composition and density may be unknown [28]. Spatial mapping is one tool that can be used to identify linkages between ecosystem services across different geographic scales, e.g., [29]. A major limitation for spatial mapping of CME services is that, unlike terrestrial ecosystems that can be mapped with remote sensing or satellite imagery, there is a scarcity of spatial data to effectively address the dynamic nature of coastal and marine environments across both spatial and temporal dimensions [19,30]. In other words, mapping what is under the surface of the ocean is much more challenging than mapping systems that can be readily seen by the naked eye. However, novel methods such as multibeam echosounder [31] and the IKONOS satellite sensor [32] are improving our capacity to map CMEs.

2.2. The Economic Valuation Process

From a valuation perspective, the basic steps to measure the TEV from CME services appear to be relatively straightforward [4,33]. First, identify the specific ecosystem service of interest and the underlying ecosystem functions and processes that produce that service. In the case of fish and shellfish provisioning from saltwater marsh and seagrass beds, for example, this would involve the biological processes (such as recruitment and growth) and the ecological requirements (such as suitable habitat and water quality) that produce these harvestable products for recreational and/or commercial users, e.g., [34]. Second, determine how existing ecological conditions contribute to production of the ecosystem services and, in the case of potential management actions or other events that would change the underlying conditions, how flows of ecosystem services could change in the future. In the context of salt marshes or seagrasses, for example, the valuation exercise would consider how changes in the area of these habitats and/or changes in salinity conditions within these habitats would impact fish and shellfish production. Third, utilize economic valuation methods to measure the TEV from the existing, and/or potential, level(s) of ecosystem services.

In general, the objective of economic valuation is to provide estimates of the net present value of ecosystem services, and a number of prior papers review the range of economic valuation methods for CMEs, e.g., [35–38]. Values for ecosystem services can be derived directly using information from prices of related market goods and services, such as by using current prices of fish products to estimate future losses in fisheries due to ocean acidification, e.g., [39]. However, most ecosystem services benefit humans in ways that are not captured in existing markets, so nonmarket valuation methods are necessary for these cases. Broadly, nonmarket valuation methods can be divided into revealed preference and stated preference methods. With revealed preference methods such as the travel cost and hedonic pricing methods, information on human behavior in existing markets, such as travel or housing, are used to estimate the value of related ecosystem services, such as beach recreation (by observing preferences for people to travel to beaches) or clean waterways (by observing price premiums for homes adjacent to clean water). In contrast, stated preference methods such as contingent valuation or discrete choice experiments rely on surveys designed to elicit people's preferences for specific ecosystem services. In recent years, studies have also used estimates of the 'social cost of carbon' to build estimates of the net present value of services related to carbon storage. The major difference between this approach and the market and nonmarket methods is that the social cost of a carbon approach does not rely on studying people's preferences, but rather it relies on simulations of the global economy under different climate change scenarios.

The specific context in which economic valuation may be utilized can significantly add to the complexity of the process. In marine planning, for example, Borger et al. [8] describe a ten-step process that begins with recognition of the initial planning authority, stakeholder participation, evaluation of existing and future conditions, and monitoring and adaptive management. In the context of the U.S. National Ocean Plan and various coastal and marine planning initiatives in the U.K., the opportunities to integrate valuation in the planning process can be very different [8]. Similarly, the European Marine Strategy Framework Directive has resulted in a mixed record of success for applications of economic valuation because of the varying requirements of the planning process [12,40].

3. Towards a Comprehensive Valuation of Coastal Ecosystem Services

3.1. Valuing the Biosphere's Services

Costanza et al.'s [41] seminal paper began a series of discussions that continue to this day on the value of the world's ecosystem services and natural capital, the feasibility of estimating these values at a global scale, and the adequacy of existing methods and data for constructing these estimates.

The original analysis relied on the use of a value or benefits transfer meta-analysis of the literature existing at the time on 17 ecosystem services across 16 biomes, which assumed a constant value per unit area for each biome. The most striking finding was an estimate of the entire biosphere's ecosystem services ranging between USD$ 16–54 trillion (10^{12}) per year, with a mean of USD$ 33 trillion per year. CMEs represented 68% of the total value, or nearly USD$ 22.4 trillion. In comparison, the study reports the value of global gross national product at the time to be USD$ 18 trillion per year. Thus, the main message was that the world's ecosystems, and CMEs alone, were more valuable than the direct market goods and services produced by the entire global economy.

At its core, this approach relies on the construction of databases of existing valuation studies that include estimated monetary value, the original units of measure, the method used to develop the estimate, the type of value estimated, the year of estimation, the original currency of the estimate, the characteristics of the ecosystem or biome, and the ecosystem service valued [42]. These databases can then be used to calculate measures such as the total or average estimated value of services produced by specific biomes or estimates of the value of ecosystem services produced per unit area for specific biomes. The databases can also be used to construct regression equations that can serve as benefit transfer functions to estimate the value of particular ecosystem services or biomes at different scales.

A series of important studies followed in the wake of Costanza et al. [41] with expansions and updates of the concept. For example, Braat et al. [43] developed the Cost of Policy Inaction (COPI) Valuation Database to quantify the global social and economic costs from the past, present, and future loss of biodiversity. They estimated that by 2050 humankind is expected to lose land-based ecosystem services worth an estimated €14 trillion, with losses in marine and coastal ecosystem services of a comparable magnitude. In contrast to the annual loss figures reported by Costanza et al. [41], the losses reported by Braat et al. [43] are cumulative losses of ecosystem services between 2008 and 2050. Similarly, de Groot et al. [42] report on progress made through the compilation and publication of The Economics of Ecosystems and Biodiversity's (TEEB) Ecosystem Service Valuation Database (ESVD), which built on the COPI database and was developed to be usable by different stakeholders to estimate values for 10 main biomes in a common spatial, temporal, and currency unit (international dollars per hectare per year). In a similar vein, Plantier-Santos et al. [44] developed the Gulf of Mexico Ecosystem Service Valuation Database (GecoServ) as an inventory of ecosystem services valuation studies applicable to the Gulf of Mexico.

Costanza et al. [6] expanded on the global approach by accounting for changes in the (per hectare per year) value of ecosystem services produced by specific biomes through time and changes in the global area covered by different biomes. The tabulation quantified the change in the value of ecosystem services between 1997 and 2011 as an annual loss ranging between USD$ 4.3–20.2 trillion. In addition, the reported values per unit area of CMEs increased between 1997 and 2011 based on new studies in the literature. However, the overall value of CMEs declined to 59.6% of the total value of global ecosystems after adjusting for both changes in unit value and areal declines, most notably in coral reefs and wetlands.

3.2. Limitations of the Global Valuation Approach

While these global-scale ecosystem valuation studies have raised public awareness about the magnitude of the contribution of ecosystems and their services to human societies, the approach has many limitations. Several researchers have pointed out that one major problem is the available data to conduct benefit transfers at a global scale are much too limited as they come from a relatively small number of primary studies over a limited geographic area. Pendleton et al. [35] were among the first to conduct a literature review on the value of CMEs. Based on over 100 studies in the U.S., they reported the valuation estimates were predominantly concentrated on four CMEs: coastal beaches, marine fishing, coastal lands, and wetlands. Also, the geographic coverage was concentrated in a few states such as Florida, California, and the Carolinas, but had scant coverage in other regions such as the Pacific Northwest, the Great Lakes, and New England.

Barbier [3] and Barbier et al. [4] provided a more global perspective on the CMEs valuation literature. These reviews reported that only coral reefs, salt marshes, and mangroves had been studied in any detail, and even these studies focused on a small subset of ecosystem services mostly related to provisioning. Barbier et al. [4] also noted a second major problem in that most studies focused on ecosystem services as independent products with little consideration of ecological interactions between services or across different ecosystems. Therefore, there is no valid basis to aggregate CME service components to determine the value of a biome if there is no understanding of the effects of relationships and feedbacks between ecosystem services (ES) components.

The Exxon Valdez oil spill in Alaska in 1989 and the Deepwater Horizon spill in the Gulf of Mexico in 2010 triggered a number of economic valuation studies focused on the impacted coastal and marine ecosystems. Most of these studies, however, relied on benefit transfers from other studies and focused mostly on provisioning services such as recreation [45]. If new research was conducted, it also focused on recreational activities, e.g., [46]. When a broad array of CME services was addressed, the analysis typically used a replacement cost approach, habitat equivalency analysis, so the results were not applicable for other areas beyond the damage zone [45]. Another focus area has been the use of valuation for coastal management decisions, particularly related to establishment of marine protected areas such as the National Marine Sanctuaries, or broadly on the value of specific ecosystems such as coral reefs or beaches. However, it has been noted that, "The variability in context, methodology, and approach in these studies do not allow for aggregation and make comparability of these resulting value estimates difficult" [45].

These prior reviews of the CME valuation literature raise concerns about the limited number of primary studies for a broad array of CME services and the lack of knowledge about how CME services may be interrelated. These two major problems can be further illustrated with the global valuation data from de Groot et al. [42] and McVittie and Hussain [47]. Table 2 presents mean and median valuation estimates for coastal and marine biomes that were used in Costanza et al. [6] and represent the existing literature at the time. Also presented in Table 2 are the percentages of the mean estimates attributed to specific categories of ecosystem services and the number of studies (in parentheses below) reported for each category.

Table 2. Ecosystem service components as a percent of total mean value (in 2007 Int.$/ha/year) for coastal and marine biomes; number of value estimates in parentheses below each percent. Adapted from: [42,47].

Biome	Total Mean (Median) Value $/ha/Year	Provisioning	Cultural	Regulating	Supporting
Coral reefs	$352,249 ($197,900)	15.8% (47)	30.9% (53)	48.9% (24)	4.6% (13)
Coastal wetlands	$193,845 ($12,163)	1.6% (80)	1.1% (20)	88.5% (30)	8.9% (12)
Other Coastal systems	$28,917 ($26,760)	8.3% (23)	1.0% (10)	89.4% (5)	1.3% (7)
Open Ocean	$491 ($135)	20.8% (7)	65.0% (5)	13.2% (1)	1.0% (3)

The valuation estimates for coral reefs in Table 2, the most studied (137 value estimates) of the CME groups, indicate it is the highest valued ($352,249 $/ha/year) with a broad range of value attributed to different ecosystem services. Regulating and supporting services comprise a majority share of the mean value estimates, although there is no way to know how an increase in provisioning services (through heavier use or harvest) would either increase or decrease the values attributed to other ecosystem services. The open ocean, on the other hand, is the least studied (16 value estimates) biome and has a mean valuation ($491 $/ha/year) that is a fraction of the value attributed to coral reefs. The vast array of ecosystem services from the open oceans (e.g., fisheries, deep water habitats, climate regulation, biodiversity, and carbon sequestration) has simply not been studied from a valuation perspective. Also, independent valuation by biome provides no recognition of the interconnectivity between open oceans and other biomes, such as coral reefs, that depend on ocean transport processes that distribute nutrients, larvae, and adult species.

For coastal wetlands and other coastal systems, the data in Table 2 indicates that the vast majority of benefits from these CMEs were derived from regulating and supporting services. There are a variety of services from these CMEs such as nutrient cycling, coastal protection (through attenuation of wave action), erosion control (through stabilization and retention of soil/organic matter and sediments), water purification (through uptake and retention of nutrients), and carbon sequestration (by fostering the accumulation of organic matter). Yet, these value estimates were based on a small number of studies, many of which were conducted more than two decades ago [42].

Criticisms of the global aggregation approach to ecosystem valuation are not new. For instance, Bockstael et al.'s [48] criticism provides an overview of the challenges inherent in a comprehensive valuation of the services provided by the biosphere's ecosystems. Economic valuation of ecosystem services entails a choice made by individuals where two alternative states of the world are weighed against each other, and each state is associated with a different level of ecosystem services being provided (however, valuing all of the world's ecosystems entails a choice where one alternative includes the biosphere as we know it, and the other entails either the absence of the biosphere's ecosystems or a state of the world that we cannot realistically define. Even if such a choice could be defined in a way that individuals could understand, their willingness (and ability) to pay to avoid such a loss of ecosystems cannot exceed their income, so the value of the world's ecosystem services cannot—by definition—exceed the combined gross income of all the people of the planet.). Pendleton et al. [49] summarized these shortcomings in a review of Costanza et al. [6] with the observation that, "despite the efforts devoted to the evaluation of the world's ecosystem services over the last three decades, these data generally are insufficient to do much more than raise awareness. In fact, the limitations of the results produced by Costanza et al. [6] illustrate the enduring lack of accurate and comprehensive, global data for ecosystem services—especially for marine and coastal areas" [49].

A major challenge in the quest for comprehensive valuation of CMEs (and ecosystem services in general) is the integration of ecological production functions into the estimation of economic valuation functions [4,33,50]. Successful valuation of changes in ecosystem services requires establishing a clear connection between an ecosystem and the ecosystem services it provides or between the ecological production function and the economic valuation function [4]. To make this connection, the changes in ecosystem structure, function, and processes that give rise to the changes in ecosystem services must be characterized. This characterization is a necessary step before a clear linkage between the changes in ecosystem structure, function, and processes, and the changes in the quantities and qualities of ecosystem service flows to people can be made. Without a clear connection, economic valuation methods are unlikely to yield meaningful and unbiased estimates.

Given that complete economic valuation of ecosystem services relies on adequate understanding of ecosystem structure, function, and processes, and their links to the flow of ecosystem services that benefit humans, valuation of benefits arising from highly complex characteristics of the natural world can be an especially challenging endeavor. A case in point is the economic valuation of biological diversity, a complex, multilevel concept that includes genetic, species, functional, molecular, and phylogenetic diversity. But, the meaning of biodiversity in political discourse and economic valuation has not been precise. Because there is no single 'right' indicator of biodiversity, economic valuation studies designed to estimate the value of biodiversity have used different proxies such as numbers of species, single species or groups of species, genetic diversity, functional diversity, habitats, or even more abstract measures such as 'low', 'medium', or 'high' biodiversity [51]. Thus, compiling value estimates from all available studies on the value of biodiversity in order to extract an average value for this complex concept is troublesome. This problem of complex processes and characteristics as valuation objects is not unique to biodiversity but a feature of supporting ecosystem services in general. Similar issues have been found in the valuation of cultural ecosystem services, where cultural ecosystem services are defined in a variety of ways [52], and researchers have recognized that the current typologies of cultural ecosystem services are insufficient to account for services such as ingenuity, life teaching, and perspective [53].

The interrelationships between multiple services produced by the same ecosystem must also be carefully considered before attempting to add up individual ecosystem service values to obtain an estimate of the ecosystem's total value. These relationships and feedbacks between ecosystem services, ecosystem components, and the ecosystems as a whole are what was previously referred to as the ecosystem production function [4,33,50]. While an ecosystem may simultaneously produce multiple services, some uses of these services may diminish the capacity of the ecosystem to produce other services. For example, use of coastal ecosystems by recreational vessels provides important cultural ecosystem services to people, but it may cause decreased water quality, introduction of non-native species, and physical disturbances [54]. Ecosystem deterioration resulting from these impacts of recreational use can be expected to alter the structure and function of the ecosystem and, thus, reduce the provision of other ecosystem services.

In the case of seagrass beds, for instance, recreational boating may lead to introduction of pathogens, pollutants, or invasive species that can disrupt the life cycle of native organisms. Similarly, high boating traffic may result in physical disturbance of the seafloor, which will result in deterioration of the sea grass beds and thus result in lower levels of regulating services such as water purification, carbon sequestration, and erosion control. Thus, there may be conflict between different types of ecosystem services where humans' use of one service may hinder production of other services. In some cases, use of cultural and provisioning services may cause deterioration or reduction in regulating and supporting services. Conversely, enhancement of a particular ecosystem service may result in complementary enhancements to other services produced by the same ecosystem. For example, restoration of oyster reefs to mitigate the effects of eutrophication also results in the creation of habitat for finfish.

Most importantly, linear changes in the areal extent of ecosystems may result in nonlinear changes to the provision of ecosystem services. For instance, linear increases in the size of mangrove forests result in quadratic increases in wave attenuation services [55]. Thus, simply adding up the values of ecosystem service components to obtain a total value for the ecosystem is likely to result in misleading estimates. Similarly, linear multiplication of areal extent of an ecosystem by an average value per area to obtain an estimate of the total value of an ecosystem relies on a strong assumption of linearity and will likely lead to biased estimates.

3.3. Current Status of Coastal and Marine Ecosystem (CME) Valuation

More recent reviews of the literature on CME valuation [18,38] show that research has continued moving in a similar direction, and thus the problems highlighted previously are still a primary concern. Torres and Hanley's [18] global review found a total of 196 studies published on CME services valuation between 2000 and 2015. The most studied CME during this period was beaches, with 40 papers published. Conversely, the least studied CME was the deep ocean, with only two papers published during this period. Most studies lacked a comprehensive approach to capture both use and nonuse values, although some recent research has integrated valuation for a broader array of CME services using stated choice valuation methods [18].

The recent literature also highlights the need for improved CME services valuation information in the study of climate change impacts on human societies and economies [56]. The value of CME services can be expected to be heavily impacted by sea level rise and extreme weather events, which will reshape the world's coastlines and the areal extent of many CMEs. In addition, warmer ocean temperatures will have significant impacts on primary productivity and the physiology of organisms inhabiting the world's oceans [34]. However, with some exceptions, e.g., [39], most of the existing approaches to estimate the future impacts of climate change rely on integrated assessment models

(IAMs) that are used to estimate a social cost of carbon (SSC) based on the expected changes in social welfare associated with different levels of carbon emissions [31,57,58]. Most of these IAMs currently ignore the nonmarket values produced by CMEs and their services [56].

Better harmonization between IAMs and nonmarket values requires a four-step procedure. First, global climate models are downscaled to a case study area where likely changes in climate can be related to physical changes in CMEs. Second, these changes are linked to specific changes in ecosystem services that are currently provided by CMEs in the study location. Third, qualitative changes in provision of affected ecosystem services are specified in quantitative terms. Last, these quantitative changes in ecosystem service provision can be valued using existing methods [38,59].

4. Valuation in Coastal and Marine Policy

Valuation of ecosystem services provided by CMEs can inform a myriad of policy decisions, including assessment of wetland mitigation strategies, design of erosion prevention and coastal conservation programs, natural resource damage assessment, fisheries management, and design of biodiversity conservation strategies, among others. In addition, valuation work can also provide important insights related to the design of financing mechanisms, including payment for environmental services schemes, as well as the implementation of environmental legislation.

4.1. Potential Uses of Valuation in the Policy Process

The use of valuation in policy and planning decisions can be defined as either informative or decisive [60]. Informative uses of ecosystem services valuation are intended to contribute to discussions and draw attention to specific resources and/or ecosystem services. For example, global studies of the total economic value of all ecosystems, e.g., [6,41] were intended principally to promote awareness and interest in the importance of ecosystem services. While these studies elevated the visibility of ecosystem services in both public and academic discourse, they relied on prior valuation studies and benefit transfer methods that lacked precision and did not address the types of problems encountered in coastal and marine ecosystem management [49,61].

Decisive uses of valuation studies in policy, on the other hand, seek to evaluate the tradeoff between different levels and/or types of ecosystem services. For example, fishery managers may set different total allowable catch limits for a coastal species that determine the amount of provisioning services from current harvests and the supporting services from future stock. Or, coastal managers may want to evaluate strategies to maintain or expand existing wetlands to sustain the storm protection benefits of wetlands and prevent future land loss from erosion and hurricanes, e.g., [62]. These tradeoffs are regional or local, and the relative benefits and costs could be compared with valuation information. A similar policy problem is the tradeoff between ecosystem services from natural capital and development. The conversion of an estuarine salt marsh to commercial or residential property will reduce the fishery production, wildlife habitat, and storm protection benefits from the salt marsh. Estimates of the value of these ecosystem services would provide a direct economic comparison of the associated benefits and costs of alternative uses for decision making.

In the following, we review some examples of the use of ecosystem services valuation in coastal and marine policy across four geographic and political regions: the European Union, the United States, Australia, and the Caribbean. We focus on the legislation or directives that may require or encourage the use of ecosystem services valuation (ESV) in planning decisions, the available literature on CME service values, and the reported success to date in the actual integration of ESVs into coastal and marine policy.

4.2. The European Union

One example of regional multinational policy setting for the use of CME services valuation is the European Marine Strategy Framework Directive established in 2008 and the Maritime Spatial Planning Directive in 2014. These directives require European Union member states to take measures to achieve or maintain "good environmental status" in the marine environment by 2020 [63]. Ecosystem-based management has been recognized as an analytical approach that establishes a broad framework to assess coastal resource and human interactions. These two directives, and other more resource-specific coastal and marine national and European legislation [63], provide for the evaluation of indicators, targets, and economic analyses to develop programs to achieve this objective and to consider transboundary features and impacts [12,40,64]. Formal guidance has been developed to utilize an ecosystem services based approach, and some results have been achieved in compiling studies and inventories for different CME service values, e.g., [65].

Despite the broad scope of the directives, there has been limited success in actual implementation of ecosystem services valuation in the planning and evaluation process. Drakou et al. [64] reviewed 11 European case studies and found that valuation studies did inform decision makers. In only a few cases, however, were the valuation results applied or used to influence decision making. The authors recommend greater involvement of 'end users' in planning, development of more integrated frameworks for coastal and marine social–ecological systems, and better informing the general public about the role of ecosystem services. Similarly, several authors [12,40,66] reported that the most common problems in the use of valuation studies involved a general lack of knowledge and expertise with ecosystem service and management concepts, gaps in understanding how changes in ecosystem functions and services impact human welfare, limited valuation data for a broad array of CME services, a lack of involvement by diverse end user and stakeholder groups in the policy planning process, and a lack of time and resources to effectively coordinate ecological and social research and planning across diverse spatial and temporal scales.

4.3. The United States

US federal government involvement in coastal and marine planning typically requires economic analyses consistent with the Water Resources Council's Principles and Guidelines established in 1983. These guidelines did not recognize ecosystem services as a component of evaluation, but the 2013 Principles, Requirements and Guidelines for Federal Investments in Water Resources [67] required an explicit recognition of ecosystem services in program management and planning [68]. In addition, the National Ocean Policy Implementation Plan [69] established a comprehensive regional approach to CME management in the US. This policy initiative recognized the linkage between CMEs and economic well-being, and it focused on expanding public awareness and data accessibility for information about the economic values associated with the broad spectrum of use and nonuse values from CME services for use in coastal and marine planning [8]. Recent revisions to the National Ocean Policy, however, may jeopardize prior directives to focus on a broad perspective for economic values from CME services in coastal and marine planning [70].

Despite these new policies, there are few reported applications of ecosystem valuation in coastal and marine resource policy and planning decision-making in the U.S. The National Centers for Coastal Ocean Science has undertaken research to estimate ecosystem values related to several estuarine reserves using a variety of valuation methods [71–73]. In addition, the issues related to state and local coastal and marine planning, where the majority of coastal planning choices occur, have been identified, and research needs have been defined, e.g., [74,75]. But, there is no published research to document whether or how ecosystem services valuation information has been used in coastal and marine policy or planning in the U.S.

4.4. Australia

In Australia, federal guidelines encourage the use of benefit–cost analyses for major regulatory programs that would impact coastal and marine resources, but actual valuation of ecosystem services is not required [66]. Despite the lack of an official policy context for valuation, considerable literature on CME services valuation for Australia has emerged [76,77]. Based on a survey of decision-makers in local, regional, state, and national government agencies as well as representatives of marine industries, Marre et al. [77] reported that ESVs were most often used in management decisions involving commercial fisheries, recreational activities and tourism, and coastal development/conservation.

Specific examples of ESV having a significant impact on decision-making were for marine park zoning in the Great Barrier Reef, e.g., [78] and protection strategies for seagrass meadows [79]. In general, however, CME service values were used mostly to communicate and support evaluation for ecosystem services. Most decision-makers believed that valuation information had a weak influence on coastal and marine policy [77]. A recent study by Sandhu et al. [80] in South Australia utilized a scenario planning process to engage a diverse group of stakeholders and decision-makers to anticipate potential future changes in economic growth, land cover, and ecosystem services from alternative coastal development pathways. While the process revealed important differences across the scenarios, there was no discussion on whether the results were integrated into regional coastal decision making.

4.5. The Caribbean

The region of the world with perhaps the greatest volume of studies on coastal and marine ecosystem valuation is the Caribbean. Schuhmann and Mahon [81] identified over 100 studies with valuation results for CME services with the majority focusing on coral reefs and marine protected areas. The quantity of research is partly attributable to efforts by the World Bank to support valuation studies in the 1990s and the Caribbean Large Marine Ecosystem project initiative by the United Nations Development Programme beginning in 2009. The latter explicitly adopted an ecosystem-based management approach to promote regional development and sustain coastal and marine ecosystem processes.

Waite et al. [82] conducted a literature review and personal interviews with coastal resource managers, government officials, and others throughout the Caribbean to determine the extent to which CME services valuation studies had been used in decision making. They reported that, in general, valuation studies had little impact on the policy process. They did identify, however, a small group of 'success stories' where the information directly influenced decision-making across several countries. The common factors that were found across these successful cases were mainly procedural and methodological and included: a clear intended use (policy question) for valuation, strong stakeholder engagement, access to decision-makers, and potential revenue enhancement for the government. Despite the limited influence from prior CME services valuation studies, the authors suggest that awareness of, and demand for, the information provided from CME services valuation is growing and could play a greater role in policy and planning in the Caribbean in the future.

5. Summary and Discussion

5.1. Summary

The development and dissemination of the ecosystem services framework in the early 2000s created a new research agenda for valuation research. The new paradigm broadened the scope of valuation concerns for coastal and marine resources from previous research that had focused almost exclusively on provisioning services such as harvesting and recreation. This new perspective gave greater recognition to a broader range of ecosystem services beyond traditional provisioning, and it put greater attention on nonuse values that had received little attention in the literature. This was especially challenging in light of the relative scarcity of data and limited understanding of spatial interconnectivity in coastal and marine ecosystems compared to terrestrial systems, e.g., [19].

This systematic review of the literature on CME services valuation indicates that studies of provisioning services still dominate valuation research. While the TEEB valuation inventory summarized in de Groot et al. [42] was notable in its attempt to expand awareness of ES components, the compilation reveals the gaping holes that exist in research and understanding of the broad range of values for CME services. It also raises serious questions about the validity of simply aggregating

ES components to determine the value of a biome when there is little understanding of the effects of relationships and feedbacks between ES components on total values.

These gaps largely reflect the fact that most CME studies took a piecemeal approach and were 'one shot' efforts for a specific problem or setting. Since recreation- and market-based activities are easier to observe and measure, use values are the most obvious to estimate. Regulating and supporting services are more obscure and typically require nonuse valuation methods. Most studies lacked a comprehensive approach to capture both use and nonuse values, although some recent research has integrated valuation for a broader array of CME services [18]. In addition, these studies rarely take into consideration the impacts that regulation and management may have on the value of interconnected provisioning and supporting ecosystem services [83].

Part of the reason for this piecemeal approach to CME services valuation is the general lack of any broader research initiatives to move beyond one-time valuation studies. As noted by several authors, e.g., [8,67,84], CMEs research and valuation studies have not been a funding priority, and policymakers oftentimes do not understand the role of ES research. Recent projects have also emerged to develop simpler, less expensive simulation tools such as InVEST, e.g., [30,85] in place of valuation research for specific coastal and marine planning problems. A major shortcoming, however, is that these policy-directed tools almost exclusively utilize benefit transfer methods and do not address issues of measurement error and validity. The valuation information from these policy tools are, on a positive note, second-best shadow values or, more likely, no more than informative indicators to recognize the role of CME services.

Despite the increased focus in the scientific community on the importance of ES, coastal and marine resource policy and planning decisions typically do not require ES valuation. Also, resource managers and decision-makers often lack a basic understanding about ES values. A review of studies from four major geographic regions across the globe indicates that CME service values are most often used for informational purposes and rarely used to evaluate tradeoffs for policy decisions. The studies cite a variety of reasons for this situation, but the most common concerns are a lack of understanding of ES for both policymakers and the public and the limited availability of valuation information for non-provisioning services in specific settings. This decision-making context for CME services policy and planning is a classic 'wicked problem' with complexity, interdependence, and conflicting social interests [22].

Some have argued that focusing on 'benefit-relevant indicators' for CME services in the planning process can provide useful information when full economic valuation is not practical or possible for the specific planning problem [86,87]. Others have argued that conventional TEV cannot account for the full range of sociocultural and ecological values from CME services. They propose a broader array of participatory and stakeholder involvement approaches for a more complete evaluation of ecosystem values in the planning process, e.g., [84,88–90].

While these alternative approaches may help in specific management settings in the short run, they do little to further a comprehensive understanding of values for a full range of CME services. Typically, these participatory approaches focus on local problems, and the participants may not reflect broader public interests, e.g., [66]. This is especially a concern in the European Union setting where the supply of ecosystem services is characterized by high degrees of spatial transboundary interdependence, e.g., [63]. Economic analysis may be difficult to apply for all CME services, but it can provide a consistent framework to develop valuation information and to evaluate that information across national boundaries.

5.2. Alternative Directions

The current approach to ES valuation for coastal and marine policy and planning is fraught with shortcomings. An alternative is to expand on the traditional national economic accounts that provide objective, regular, and standardized information that public and private interests rely on for planning and decision making [91]. A broader system of 'wealth accounting' would include natural capital accounts (NCAs) and ecosystem goods and service accounts (EGSAs) that could provide governments and businesses with information on the current status of a nation's (or the entire biosphere's) natural capital and ecosystem services [23,92,93]. This wealth accounting framework is not new, e.g., [94,95], but the ecosystem services paradigm has renewed international attention on the merits of a comprehensive approach to environmental valuation [24]. This alternative also does not negate the need for, and importance of, smaller scale valuation studies that link ecological production functions to specific economic benefits for coastal and marine policy decision making, e.g., [81].

For coastal and marine policy and planning, adopting the wealth accounting framework is an ambitious undertaking. But, the conceptual and practical frameworks have been defined. For example, a wealth accounting framework would include changes in fishery stocks as part of national income in much the same way as changes in farmed animal stocks are now included. The reason why fish stocks are excluded is because they are considered natural assets beyond the production boundary of the current system, whereas livestock is a produced asset controlled by human activities. Guidelines for implementing this approach for fisheries have existed for many years, e.g., [96], and additional guidance to expand the framework for a broad range of natural assets and ecosystem services is also available, e.g., [16].

The primary advantage of the wealth accounting framework is that it provides consistent, internationally recognized standards for valuation [17,91]. A major challenge for applying this framework to coastal and marine resources is that many of the ecosystem services are nonmarket outputs and often provide nonuse values. Provisioning and cultural services are fully amenable to wealth accounting, but regulating and supporting services must be tracked through to the final consumption of households distinct from market goods. This requires a separate accounting of nonmarket services consumption that can highlight the relative contribution of different ecosystem services to human welfare [24,97].

The information that can be available for coastal and marine policy and planning from a wealth accounting approach is important to inform the public about the relative importance of CME services and to evaluate tradeoffs between different types of CME services. For example, Barbier [92] demonstrates how households' willingness to pay for restoration of estuarine and coastal habitats is directly related to the expected future storm protection benefits from these restored habitats. Similarly, if extraction activities to harvest the provisioning services available in the short run from a CME threaten sustainability of a basket of other supporting and regulating services in the long run, then the total value of the CME will be degraded in the future. A coral reef, for example, that provides short-term benefits from the provision of recreation and harvesting but is over-utilized or degraded due to human visitation, may experience reduced levels of fishery habitat and biodiversity services in the future. A comprehensive valuation framework is needed that recognizes these tradeoffs and the capacity of the CME to generate these ecosystem services in the future. Successful implementation of a wealth accounting framework will require a commitment by national governments across the world to fully inform the public about the important services provided by natural capital such as CMEs.

The ecosystem services paradigm has changed the terms of debate about coastal and marine valuation for policy and planning. There have been clear inroads in public recognition of the role of ecosystem services, and there are some examples around the world where CME services valuation information has had an impact on policy and decision making. The question looking forward is whether this new ecosystem services perspective can change the direction of research on CME services valuation and the information available to decision-makers and the public.

Author Contributions: Conceptualization, J.W.M.; methodology, J.W.M.; software, J.W.M., S.A.; validation, J.W.M., S.A.; formal analysis, J.W.M., S.A.; investigation, J.W.M.; resources, J.W.M., S.A.; data curation, J.W.M., S.A.; writing—original draft preparation, J.W.M., S.A.; writing—review and editing, J.W.M., S.A.; visualization, J.W.M., S.A.; supervision, J.W.M.; project administration, J.W.M.

Funding: This research received no external funding.

Conflicts of Interest: The authors declare no conflict of interest.

References

1. Neumann, B.; Vafeidis, A.; Zimmermann, J.; Nicholls, R. Future coastal population growth and exposure to sea-level rise and coastal flooding-a global assessment. *PLoS ONE* **2015**, *10*, e0118571. [CrossRef] [PubMed]
2. Millennium Ecosystem Assessment. *Ecosystems and Human Well-Being: Synthesis*; Island Press: Washington, DC, USA, 2005.
3. Barbier, E.B. Progress and challenges in valuing coastal and marine ecosystem services. *Rev. Environ. Econ. Policy* **2012**, *6*, 1–19. [CrossRef]
4. Barbier, E.B.; Hacker, S.; Kennedy, C.; Koch, E.; Stier, A.; Sillman, B. The value of estuarine and coastal ecosystem services. *Ecol. Monogr.* **2011**, *81*, 169–193. [CrossRef]
5. Halpern, B.S.; Walbridge, S.; Selkoe, K.A.; Kappel, C.V.; Micheli, F.; D'agrosa, C.; Bruno, J.F.; Casey, K.S.; Ebert, C.; Fox, H.E.; et al. A global map of human impact on marine ecosystems. *Science* **2008**, *319*, 948–952. [CrossRef] [PubMed]
6. Costanza, R.; de Groot, R.; Sutton, P.; Van der Ploeg, S.; Anderson, S.J.; Kubiszewski, I.; Farber, S.; Turner, R.K. Changes in the global value of ecosystem services. *Glob. Environ. Chang.* **2014**, *26*, 152–158. [CrossRef]
7. Fourqurean, J.; Duarte, C.; Kennedy, H.; Marbà, N.; Holmer, M.; Mateo, M.; Apostolaki, E.; Kendrick, G.; Krause-Jensen, D.; McGlathery, K.; et al. Seagrass ecosystems as a globally significant carbon stock. *Nat. Geosci.* **2012**, *5*, 505–509. [CrossRef]
8. Borger, T.; Beaumont, N.; Pendleton, L.; Boyle, K.; Cooper, P.; Fletcher, S.; Haab, T.; Hanemann, M.; Hooper, T.; Hussain, S.; et al. Incorporating ecosystem services in marine planning: The role of valuation. *Mar. Policy* **2014**, *46*, 161–170. [CrossRef]
9. US Environmental Protection Agency. *Chesapeake Bay Total Maximum Daily Loads for Nitrogen, Phosphorous and Sediment*; US Environmental Protection Agency: Washington, DC, USA, 2010. Available online: https://www.epa.gov/chesapeake-bay-tmdl/chesapeake-bay-tmdl-document (accessed on 10 May 2019).
10. US National Estuary Program. *National Estuary Program: Program Evaluation Guidance*; US Environmental Protection Agency: Washington, DC, USA, 2016. Available online: https://www.epa.gov/nep/progress-evaluation-national-estuary-program (accessed on 10 May 2019).
11. US Environmental Protection Agency. *The Gulf of Mexico Program: Protecting and Preserving the Gulf of Mexico*; 2018 Annual Report; US Environmental Protection Agency: Washington, DC, USA, 2018. Available online: https://www.epa.gov/sites/production/files/2018-12/documents/2018_gulf_of_mexico_program_annual_report.pdf (accessed on 10 May 2019).
12. Van der Veeren, R.; Buchs, A.; Hormandinger, G.; Oinonen, S.; Santos, C. Ten years of economic analyses for the European Marine Strategy Framework Directive: Overview of experiences and lessons. *J. Ocean Coast. Econ.* **2018**, *5*, 1. [CrossRef]
13. US Environmental Protection Agency. *Valuing the Protection of Ecological Systems and Services*; EPA Science Advisory Panel: Washington, DC, USA, 2009.
14. US Environmental Protection Agency. *National Ecosystem Services Classification System: Framework Design and Policy Application. EPA-800-R-15-002*; Environmental Protection Agency: Washington, DC, USA, 2015.
15. Haines-Young, R. *Report of Results of a Survey to Assess the Use of the Common International Classification of Ecosystem Services*; European Environment Agency: Copenhagen, Denmark, 2016.
16. United Nations. *System of Environmental-Economic Accounting 2012, Experimental Ecosystem Accounting*; United Nations: New York, NY, USA, 2014.
17. International Organization for Standardization. *ISO 14008:2019, Monetary Valuation of Environmental Impacts and Related Environmental Aspects*; International Organization for Standardization: Geneva, Switzerland, 2019.
18. Torres, C.; Hanley, N. Communicating research on the economic valuation of coastal and marine ecosystem services. *Mar. Policy* **2017**, *75*, 99–107. [CrossRef]
19. Townsend, M.; Davies, K.; Hanley, N.; Hewitt, J.; Lundquist, C.; Lohrer, A. The challenge of implementing the marine ecosystem service concept. *Front. Mar. Sci.* **2018**, *5*, 1–13. [CrossRef]
20. Fisher, B.; Turner, K.; Zylstra, M.; Brouwer, R.; De Groot, R.; Farber, S.; Ferraro, P.; Green, R.; Hadley, D.; Harlow, J.; et al. Ecosystem services and economic theory: Integration for policy-relevant research. *Ecol. Appl.* **2008**, *18*, 2050–2067. [CrossRef] [PubMed]

21. Chan, K.M.; Balvanera, P.; Benessaiah, K.; Chapman, M.; Díaz, S.; Gómez-Baggethun, E.; Gould, R.; Hannahs, N.; Jax, K.; Klain, S.; et al. Why protect nature? Rethinking values and the environment. *Proc. Natl. Acad. Sci. USA* **2016**, *113*, 1462–1465. [CrossRef] [PubMed]

22. Davies, K.; Fisher, K.; Dickson, M.; Thrush, S.; Le Heron, R. Improving ecosystem service frameworks to address wicked problems. *Ecol. Soc.* **2015**, *20*, 37. [CrossRef]

23. Barbier, E. Wealth accounting, ecological capital and ecosystem services. *Environ. Dev. Econ.* **2013**, *18*, 133–161. [CrossRef]

24. Obst, C.; Hein, L.; Edens, B. National accounting and the valuation of ecosystem assets and their services. *Environ. Resour. Econ.* **2015**, *64*, 1–23. [CrossRef]

25. Arrow, K.J.; Cropper, M.L.; Eads, G.C.; Hahn, R.W.; Lave, L.B.; Noll, R.G.; Portney, P.R.; Russell, M.; Schmalensee, R.; Smith, V.K.; et al. Is there a role for benefit-cost analysis in environmental, health, and safety regulation? *Science* **1996**, *272*, 221–222. [CrossRef] [PubMed]

26. Davidson, M. On the relation between ecosystem services, intrinsic value, existence value and economic valuation. *Ecol. Econ.* **2013**, *95*, 171–177. [CrossRef]

27. Drakou, E.; Pendleton, L.; Effron, M.; Ingram, J.; Teneva, L. When ecosystems and their services are not co-located: Oceans and coasts. *ICES J. Mar. Sci.* **2017**, *74*, 1531–1539. [CrossRef]

28. Hewitt, J.; Ellis, J.; Thrush, S. Multiple stressors, nonlinear effects and the implications of climate change impacts on marine coastal ecosystems. *Glob. Chang. Biol.* **2016**, *22*, 2665–2675. [CrossRef]

29. Martinez-Harms, M.J.; Balvanera, P. Methods for mapping ecosystem service supply: A review. *Int. J. Biodivers. Sci. Ecosyst. Serv. Manag.* **2012**, *8*, 17–25. [CrossRef]

30. Guerry, A.D.; Ruckelshaus, M.H.; Arkema, K.K.; Bernhardt, J.R.; Guannel, G.; Kim, C.K.; Marsik, M.; Papenfus, M.; Toft, J.E.; Verutes, G.; et al. Modeling benefits from nature: Using ecosystem services to inform coastal and marine spatial planning. *Int. J. Biodivers. Sci. Ecosyst. Serv. Manag.* **2012**, *8*, 107–121. [CrossRef]

31. Tyllianakis, E.; Callaway, A.; Vanstaen, K.; Luisetti, T. The value of information: Realizing the economic benefits of mapping seagrass meadows in the British Virgin Islands. *Sci. Total Environ.* **2019**, *650*, 2107–2116. [CrossRef] [PubMed]

32. Knudby, A.; Nordlund, L. Remote sensing of seagrasses in a patchy multi-species environment. *Int. J. Remote Sens.* **2011**, *32*, 2227–2244. [CrossRef]

33. Polasky, S.; Segerson, K. Integrating ecology and economics in the study of ecosystem services: Some lessons learned. *Ann. Rev. Resour. Econ.* **2009**, *1*, 409–434. [CrossRef]

34. Bell, F. The economic valuation of saltwater marsh supporting marine recreational fishing in the southeastern United States. *Ecol. Econ.* **1997**, *21*, 243–254. [CrossRef]

35. Pendleton, L.; Atiyah, P.; Moorthy, A. Is the non-market literature adequate to support coastal and marine management? *Ocean Coast. Manag.* **2007**, *50*, 363–378. [CrossRef]

36. Waite, R.; Burke, L.; Gray, E.; van Beukering, P.; Brander, L.; McKenzie, E.; Pendleton, L.; Schuhmann, P.; Tompkins, E. *Coastal Capital: Ecosystem Valuation for Decision Making in the Caribbean*; World Resources Institute: Washington, DC, USA, 2014.

37. Vassilopoulos, A.; Koundouri, P. *Valuation of Marine Ecosystems*; Oxford Research Encyclopedia of Environmental Science: Oxford, UK, 2017.

38. Mehvar, S.; Filatova, T.; Dastgheib, A.; de Ruyter van Steveninck, E.; Ranasinghe, R. Quantifying economic value of coastal ecosystem services: A review. *J. Mar. Sci. Eng.* **2018**, *6*, 5. [CrossRef]

39. Mangi, S.; Lee, J.; Pinnegar, J.; Law, R.; Tyllianakis, E.; Birchenough, S. The economic impacts of ocean adification on shellfish and aquaculture in the United Kingdom. *Environ. Sci. Policy* **2018**, *86*, 95–105. [CrossRef]

40. Hanley, N.; Hynes, S.; Patterson, D.; Jobstvogt, N. Economic valuation of marine and coastal ecosystems: Is it currently fit for purpose? *J. Ocean Coast. Econ.* **2015**, *2*, 1. [CrossRef]

41. Costanza, R.; d'Arge, R.; de Groot, R.; Farber, S.; Grasso, M.; Hannon, B.; Limburg, K.; Naeem, S.; O'Neill, R.V.; Paruelo, J.; et al. The value of the world's ecosystem services and natural capital. *Nature* **1997**, *387*, 253–260. [CrossRef]

42. De Groot, R.; Brander, L.; Van Der Ploeg, S.; Costanza, R.; Bernard, F.; Braat, L.; Christie, M.; Crossman, N.; Ghermandi, A.; Hein, L.; et al. Global estimates of the value of ecosystems and their services in monetary units. *Ecosyst. Serv.* **2012**, *1*, 50–61. [CrossRef]

43. Braat, L.; Ten Brink, P.; Bakkes, J.; Bolt, K.; Braeuer, I.; Ten Brink, B.; Chiabai, A.; Ding, H.; Gerdes, H.; Jeuken, M.; et al. *The Cost of Policy Inaction (COPI): The Case of Not Meeting the 2010 Biodiversity Target*. 2008. Available online: https://www.cbd.int/financial/doc/copi-2008.pdf (accessed on 22 July 2019).

44. Plantier-Santos, C.; Carollo, C.; Yoskowitz, D. Gulf of Mexico ecosystem service valuation database (GecoServ): Gathering ecosystem services valuation studies to promote their inclusion in the decision-making process. *Mar. Policy* **2012**, *36*, 214–217. [CrossRef]

45. Lipton, D.; Lew, D.; Wallmo, K.; Wiley, P.; Dvarskas, A. The evolution of non-market valuation of U.S. coastal and marine resources. *J. Ocean Coast. Econ.* **2014**, *2014*, 6. [CrossRef]

46. Alvarez, S.; Larkin, S.L.; Whitehead, J.C.; Haab, T. A revealed preference approach to valuing non-market recreational fishing losses from the Deepwater Horizon oil spill. *J. Environ. Manag.* **2014**, *145*, 199–209. [CrossRef]

47. McVittie, A.; Hussain, D. *The Economics of Ecosystems and Biodiversity—Valuation Database Manual*; Scotland Rural College: Edinburgh, Scotland, 2013.

48. Bockstael, N.E.; Freeman, A.M.; Kopp, R.J.; Portney, P.R.; Smith, V.K. On measuring economic values for nature. *Environ. Sci. Technol.* **2000**, *34*, 1384–1389. [CrossRef]

49. Pendleton, L.; Thébaud, O.; Mongruel, R.; Levrel, H. Has the value of global marine and coastal ecosystem services changed? *Mar. Policy* **2016**, *64*, 156–158. [CrossRef]

50. National Research Council. *Valuing Ecosystem Services: Toward Better Environmental Decision Making'*; National Academies Press: Washington, DC, USA, 2005.

51. Bartkowski, B.; Lienhoop, N.; Hansjurgens, B. Capturing the complexity of biodiversity: A critical review of economic valuation studies of biological diversity. *Ecol. Econ.* **2015**, *113*, 1–14. [CrossRef]

52. Milcu, A.; Hanspach, J.; Abson, D.; Fischer, J. Cultural ecosystem services: A literature review and prospects for future research. *Ecol. Soc.* **2013**, *18*, 44. [CrossRef]

53. Gould, R.; Lincoln, N. Expanding the suite of cultural ecosystem services to include ingenuity, perspective, and life teaching. *Ecosyst. Serv.* **2017**, *25*, 117–127. [CrossRef]

54. Hardiman, N.; Burgin, S. Recreational impacts on the fauna of Australian coastal marine ecosystems. *J. Environ. Manag.* **2010**, *91*, 2096–2108. [CrossRef]

55. Barbier, E.B.; Koch, E.W.; Silliman, B.R.; Hacker, S.D.; Wolanski, E.; Primavera, J.; Granek, E.F.; Polasky, S.; Aswani, S.; Cramer, L.A.; et al. Coastal ecosystem-based management with nonlinear ecological functions and values. *Science* **2008**, *319*, 321–323. [CrossRef] [PubMed]

56. Burke, M.; Craxton, M.; Kolstad, C.D.; Onda, C.; Allcott, H.; Baker, E.; Barrage, L.; Carson, R.; Gillingham, K.; Graff-Zivin, J.; et al. Opportunities for advances in climate change economics. *Science* **2016**, *352*, 292–293. [CrossRef] [PubMed]

57. Luisetti, T.; Jackson, E.; Turner, R. Valuing the European 'coastal blue carbon' storage benefit. *Mar. Pollut. Bull.* **2013**, *71*, 101–106. [CrossRef] [PubMed]

58. Luisetti, T.; Turner, K.; Andrews, J.; Jickells, D.; Kroger, S.; Diesing, M.; Paltriguera, L.; Johnson, M.; Parker, E.; Bakker, D.; et al. Quantifying and valuing carbon flows and stores in coastal and shelf ecosystems in the UK. *Ecosyst. Serv.* **2019**, *35*, 67–76. [CrossRef]

59. Walsh, P.; Griffiths, C.; Guignet, D.; Klemick, H. Adaptation, sea level rise, and property prices in the Chesapeake Bay Watershed. *Land Econ.* **2019**, *95*, 19–34. [CrossRef] [PubMed]

60. Laurans, Y.; Rankovic, A.; Billé, R.; Pirard, R.; Mermet, L. Use of ecosystem services economic valuation for decision making: Questioning a literature blindspot. *J. Environ. Manag.* **2013**, *119*, 208–219. [CrossRef]

61. Nelson, J.; Kennedy, P. The use (and abuse) of meta-analyses in environmental and natural resource economics: An assessment. *Environ. Resour. Econ.* **2009**, *42*, 345–377. [CrossRef]

62. Petrolia, D.; Kim, T. Preventing land loss in coastal Louisiana: Estimates of WTP and WTA. *J. Environ. Manag.* **2011**, *92*, 859–865. [CrossRef]

63. Cavallo, M.; Borja, A.; Elliott, M.; Quintino, V.; Touza, J. Impediments to achieving integrated marine management across borders: The case of the EU Marine Strategy Framework Directive. *Mar. Policy* **2019**, *109*, 68–73. [CrossRef]

64. Drakou, E.G.; Kermagoret, C.; Liquete, C.; Ruiz-Frau, A.; Burkhard, K.; Lillebø, A.I.; van Oudenhoven, A.P.; Ballé-Béganton, J.; Rodrigues, J.G.; Nieminen, E.; et al. Marine and coastal ecosystem services on the science–policy–practice nexus: Challenges and opportunities from 11 European case studies. *Int. J. Biodivers. Sci. Ecosyst. Serv. Manag.* **2018**, *3*, 51–67. [CrossRef]

65. Skourtos, M.; Damigos, D.; Tsitakis, D.; Koontogianni, A.; Tourkolias, C.; Streftaris, N. In search of marine ecosystem service values: The V-MESSES database. *J. Environ. Assess. Policy Manag.* **2015**, *17*, 4. [CrossRef]

66. Dunford, R.; Harrison, P.; Smith, A.; Dick, J.; Barton, D.N.; Martin-Lopez, B.; Kelemen, E.; Jacobs, S.; Saarikoski, H.; Turkelboom, F. Integrating methods for ecosystem service assessment: Experiences from real world situations. *Ecosyst. Serv.* **2018**, *29*, 499–514. [CrossRef]

67. Updated Principles, Requirements and Guidelines for Water and Land Related Resources Implementation Studies. Available online: https://obamawhitehouse.archives.gov/administration/eop/ceq/initiatives/PandG (accessed on 21 May 2019).

68. Schaefer, M.; Goldman, E.; Bartuska, A.M.; Sutton-Grier, A.; Lubchenco, J. Nature as capital: Advancing and incorporating ecosystem services in United States federal policies and programs. *Proc. Natl. Acad. Sci. USA* **2015**, *112*, 7383–7389. [CrossRef]

69. National Ocean Council. *National Ocean Policy Implementation Plan*; Office of the President: Washington, DC, USA, 2013.

70. Malakoff, D. Trump's new oceans policy washes away Obama's emphasis on conservation and climate. *Sci. Mag.* **2018**. Available online: https://www.sciencemag.org/news/2018/06/trump-s-new-oceans-policy-washes-away-obama-s-emphasis-conservation-and-climate (accessed on 18 July 2019).

71. Loerzel, J.; Gorstein, M.; Rezaie, A.M.; Gonyo, S.B.; Fleming, C.S.; Orthmeyer, A. *Economic Valuation of Shoreline Protection within the Jacques Cousteau National Estuarine Research Reserve*; NOAA Technical Memorandum NOS NCCOS 234: Silver Spring, MD, USA, 2017; p. 78. Available online: https://repository.library.noaa.gov/view/noaa/16081 (accessed on 22 July 2019).

72. Loerzel, J.; Knapp, L.; Gorstein, M. *Gauging the Social Values of Ecosystem Services in the Mission-Aransas National Estuarine Research Reserve*; NOAA Technical Memorandum NOS NCCOS 243: Silver Spring, MD USA, 2017; p. 79. Available online: https://repository.library.noaa.gov/view/noaa/17250 (accessed on 22 July 2019).

73. Loerzel, J.; Fleming, C.S.; Gorstein, M. *Ecosystem Services Valuation of the Central Georgia Coast, including Sapelo Island National Estuarine Research Reserve and Gray's Reef National Marine Sanctuary*; NOAA Technical Memorandum NOS NCCOS 248: Silver Spring, MD, USA, 2018; p. 85.

74. Eastern Research Group, Inc. *A Policy Analysis of the use of Ecosystem Service Values on State and Local Decision-Making: Potential Policy Questions and Gaps Analysis*; NOAA Coastal Services Center: Charleston, SC, USA, 2014.

75. National Science and Technology Council, Committee on Environment, Natural Resources, and Sustainability. *Ecosystem-Service Assessment: Research Needs for Coastal Green Infrastructure*; National Science and Technology Council, Committee on Environment, Natural Resources, and Sustainability: Washington, DC, USA, 2015.

76. Rogers, A.A.; Kragt, M.E.; Gibson, F.L.; Burton, M.P.; Petersen, E.H.; Pannell, D.J. Non-market valuation: Usage and impacts in environmental policy and management in Australia. *Aust. J. Agric. Resour. Econ.* **2013**, *59*, 1–15. [CrossRef]

77. Marre, J.B.; Thébaud, O.; Pascoe, S.; Jennings, S.; Boncoeur, J.; Coglan, L. Is economic valuation of ecosystem services useful to decision-makers? Lessons learned from Australian coastal and marine management. *J. Environ. Manag.* **2016**, *178*, 52–62. [CrossRef] [PubMed]

78. Stoeckl, N.; Hicks, C.C.; Mills, M.; Fabricius, K.; Esparon, M.; Kroon, F.; Kaur, K.; Costanza, R. The economic value of ecosystem services in the Great Barrier Reef: Our state of knowledge. *Ann. N. Y. Acad. Sci.* **1219**, *1*, 113–133. [CrossRef] [PubMed]

79. Deans, J.; Murray-Jones, S. Value of seagrass meadows to the metropolitan Adelaide coast. In *Proceedings of the Coasts and Ports 2003 Australasian Conference*; Kench, P., Hume, T., Eds.; Institution of Engineers, Australia: Barton, Australia, 2003. Available online: https://search.informit.com.au/documentSummary;dn=213137853646436;res=IELENG;type=pdf (accessed on 22 July 2019).

80. Sandhu, H.; Clarke, B.; Baring, R.; Anderson, S.; Fisk, C.; Dittman, S.; Walker, S.; Sutton, P.; Kubiszewski, I.; Costanza, R. Scenario planning including ecosystem services for a coastal region in South Australia. *Ecosyst. Serv.* **2018**, *31*, 194–207. [CrossRef]

81. Schuhmann, P.; Mahon, R. The valuation of marine ecosystem goods and services in the Caribbean: A literature review and framework for future valuation efforts. *Ecosyst. Serv.* **2015**, *11*, 56–66. [CrossRef]

82. Waite, R.; Kushner, B.; Jungwiwattanaporn, M.; Gray, E.; Burke, L. Use of coastal economic valuation in decision making in the Caribbean: Enabling conditions and lessons learned. *Ecosyst. Serv.* **2015**, *11*, 45–55. [CrossRef]

83. Barbier, E.B. Valuing coastal habitat-fishery linkages under regulated open access. *Water* **2019**, *11*, 847. [CrossRef]

84. Pandeya, B.; Buytaert, W.; Zulkaflie, T.; Karpouzoglou, T.; Mao, F.; Hannah, D. A comparative analysis of ecosystems services valuation approaches for application at the local scale and in data scarce regions. *Ecosyst. Serv.* **2016**, *22*, 250–259. [CrossRef]

85. Ruckelshaus, M.; McKenzie, E.; Tallis, H.; Guerry, A.; Daily, G.; Kareiva, P.; Polasky, S.; Ricketts, T.; Bhagabati, N.; Wood, S.A.; et al. Notes from the field: Lessons learned from using ecosystem service approaches to inform real-world decisions. *Ecol. Econ.* **2015**, *115*, 11–21. [CrossRef]

86. Olander, L.; Polasky, S.; Kagan, J.S.; Johnston, R.J.; Wainger, L.; Saah, D.; Maguire, L.; Boyd, J.; Yoskowitz, D. So you want your research to be relevant? Building the bridge between ecosystem services research and practice. *Ecosyst. Serv.* **2017**, *26*, 170–182. [CrossRef]

87. Wainger, L.A.; Secor, D.H.; Gurbisz, C.; Kemp, W.M.; Gilbert, P.M.; Houde, E.D.; Richkus, J.; Barber, M.C. Resilience indicators support valuation of estuarine ecosystem restoration under climate change. *Ecosyst. Health Sustain.* **2017**, *3*, e01268. [CrossRef]

88. Lopes, R.; Videira, N. Valuing marine and coastal ecosystem services: An integrated participatory framework. *Ocean Coast. Manag.* **2013**, *84*, 153–162. [CrossRef]

89. Jacobs, S.; Martín-López, B.; Barton, D.N.; Dunford, R.; Harrison, P.A.; Kelemen, E.; Saarikoski, H.; Termansen, M.; García-Llorente, M.; Gómez-Baggethun, E.; et al. The means determine the end—Pursuing integrated valuation in practice. *Ecosyst. Serv.* **2018**, *29*, 515–528. [CrossRef]

90. Pendleton, L.; Mongruel, R.; Beaumont, N.; Hooper, T.; Charles, M. A triage approach to improve the relevance of marine ecosystem services assessments. *Mar. Ecol. Prog. Ser.* **2015**, *530*, 183–193. [CrossRef]

91. Boyd, J.W.; Bagstad, K.J.; Ingram, J.C.; Shapiro, C.D.; Adkins, J.E.; Casey, C.F.; Duke, C.S.; Glynn, P.D.; Goldman, E.; Grasso, M.; et al. The natural capital accounting opportunity: Let's really do the numbers. *Bioscience* **2018**, *68*, 940–943. [CrossRef]

92. Dasgupta, P. The welfare economic theory of green national accounts. *Environ. Resour. Econ.* **2009**, *42*, 3–38. [CrossRef]

93. Barbier, E. The protective value of estuarine and coastal ecosystem services in a wealth accounting framework. *Environ. Resour. Econ.* **2016**, *64*, 37–58. [CrossRef]

94. Ahmad, Y.J.; El Serafy, S.; Lutz, E. *Environmental Accounting for Sustainable Development: A UNEP World Bank Symposium*; The World Bank: Washington, DC, USA, 1989.

95. Bartelmus, P. Toward a system of integrated environmental and economic accounts (SEEA). In *Integrating Economic and Ecological Indicators*; Milon, J., Shogren, J., Eds.; Praeger: Westport, CT, USA, 1995.

96. United Nations Food and Agriculture Organization. *Integrated Environmental and Economic Accounting for Fisheries. Handbook of National Accounting*; United Nations Food and Agriculture Organization: Rome, Italy, 2004.

97. Schroter, M.; Barton, D.N.; Remme, R.P.; Hein, L. Accounting for capacity and flow of ecosystem services: A conceptual model and a case study for Telemark, Norway. *Ecol. Indic.* **2014**, *36*, 539–551. [CrossRef]

water

MDPI

Article

Valuing Coastal Habitat–Fishery Linkages under Regulated Open Access

Edward B. Barbier

Department of Economics, Colorado State University, Fort Collins, CO 80523-1771, USA;
Edward.barbier@colostate.edu; Tel.: +970-491-6324

Received: 28 February 2019; Accepted: 10 April 2019; Published: 23 April 2019

Abstract: This paper explores how regulation of an open access fishery influences the value of a coastal habitat that serves as breeding and nursery grounds. A model of the fishery supported by a coastal wetland is developed, which includes a quota rule that restricts harvest to a fixed proportion of the current stock. The model is applied to mangrove-dependent shellfish and demersal fisheries in Thailand. The value of the welfare effects associated with a change in a supporting coastal habitat is influenced significantly by whether or not the regulatory quota can adjust in response to these changes. Welfare losses are considerably higher when the quota is fixed as opposed to when it can be adjusted. With the restriction in place, effort cannot change to offset the decline in biomass, and as a result, there is a much larger fall in harvest. In addition, the welfare losses are much larger for the shellfish compared to the demersal fisheries. The analysis illustrates that imposing a regulatory rule on an open access fishery has important implications for valuing any linkage between coastal breeding and nursery habitat and a near-shore fishery.

Keywords: economic valuation; estuarine and coastal ecosystems; fishery; habitat–fishery linkages; mangroves; open access; regulated open access; quota; Thailand; wetlands

1. Introduction

Estuarine and coastal ecosystems worldwide provide a wide variety of important and valuable ecosystem services, from providing food and raw materials to protecting against storms, flooding and coastal erosion, to providing habitat for marine species and biodiversity [1–3]. An important benefit of many estuarine and coastal ecosystems is their role in providing breeding and nursery habitats that support near-shore fisheries. This has resulted in the development of bioeconomic models to capture this effect and value coastal-habitat linkages (see [1,4] for reviews).

Beginning with [5], studies have shown that the values estimated for improved coastal and marine habitat quality for fisheries, whether through protection of the nursery and breeding habitat service of coastal wetlands or reduction of nutrient pollution, vary with open access versus optimal management conditions [5–8]. Other studies have pointed to the influence of market structure and the regulatory environment on fisheries that are threatened by disease and other more general environmental risks [9,10].

Previous work has demonstrated how open access management conditions prevalent in many fisheries influence the value imputed to coastal habitat changes [5–7]. First, if an open access fishery is more heavily exploited in the long run, the subsequent economic losses associated with the destruction of natural habitat supporting this fishery are likely to be lower. Second, the welfare effects associated with a change in a supporting coastal habitat will vary significantly with the magnitude of the elasticity of demand for the harvested fish. Thus, one should be cautious about basing coastal development decisions solely on single services such as habitat-fishery linkages when open access conditions may distort the impact of conserving coastal wetlands on increased fishery production and returns.

In comparison, there has been a lack of analysis of how regulation of an open access fishery may influence the value imputed to coastal habitat-fishery linkages. Regulated open access refers to the imposition of biologically motivated regulations on fishers, who would otherwise enter and exit harvesting at will [11]. A number of modeling approaches have explored the economic implications for regulating an open access fishery, based on limiting the total catch through restricting entry and season length [11–14]. A common assumption is that the allowable quota on harvesting is set relative to the biomass of the fish population [13]. The models predict that regulating an open access fishery significantly impacts harvest and biomass levels, thus affecting costs and revenues. Since valuing coastal-habitat-fishery linkages is also measured through its impact on costs and economic returns, this implies that any regulation imposed on an open access fishery would influence the value considerably.

Two studies of how regulation impacts a fishery may offer some clues on what could matter in valuing coastal-habitat linkages under regulated open access [10,12]. For example, Deacon et al. [12] explore how a regulatory rule amounts to a limit on capital, but fishing firms can vary any unrestricted input, and thereby use the restricted input more intensively. In effect, this is analogous to the problem faced by a fishery when the coastal habitat supporting it declines. One input into the fishery—the biomass stock—will decrease, or essentially become 'restricted', in the long run. Although the second input, fishing effort, may vary in response to the declining stock, this could depend on whether or not the regulatory rule is fixed and restricts effort. Kennedy and Barbier [10] show that such "target flexibility" in response to environmental shocks on a regulated fishery is critical. When the shocks impact both costs and growth, harvest quotas are strongly favored over both effort quotas and taxes, but losses are reduced by harvest control mechanisms that can be adjusted in response to these shocks. As a declining coastal breeding and nursery habitat is also likely to impact costs and growth of a fishery, then valuing any changes in the habitat-fishery linkage could depend on whether or not the regulatory regime is adjusted in response to the habitat loss.

To summarize, the literature on valuing coastal-habitat-fishery linkages under open access suggests that the magnitude of the elasticity of demand is an important determinant in the welfare effects associated with a change in a supporting coastal habitat. In contrast, models on regulated open access indicate that restricting harvests so that they do not over-exploit the fish stock is likely to have implications for biomass size, costs, and returns, and thus may impact significantly the value of coastal-habitat-ishery linkages. In addition, the losses associated with any decline in coastal habitat linkages will depend on whether or not there is flexibility in adjusting the regulatory quota on harvest.

The purpose of the following paper is to explore these influences on valuing habitat-fishery linkages in a regulated open access fishery. A model of the fishery supported by a coastal wetland is developed. A quota rule is imposed to ensure that harvest must always be a fixed proportion of the current stock. The key relationships are kept analytically tractable in order to show as explicitly as possible how the regulatory restriction, compared to the elasticity of market demand, impacts the value of any change in the coastal habitat supporting the fishery. The model also indicates how the value of this habitat-fishery linkage will depend on whether or not the regulatory quota is adjusted in response to a decline in the wetland area.

Further insights into these relationships are gained by an empirical application based on mangrove-dependent shellfish and demersal fisheries in Thailand, from data and parameters estimated in [6]. Both the theoretical model and its simulation indicate that the elasticity of demand does impact the value of the welfare effects associated with a change in a supporting coastal habitat. However, this value is influenced even more by whether or not the regulatory quota can adjust in response to these changes. Welfare losses are considerably higher when the quota is fixed as opposed to when it can be adjusted. With the restriction in place, effort cannot change to offset the decline in biomass, and as a result, there is a much larger fall in harvest. In addition, the welfare losses are much more significant for the shellfish compared to the demersal fisheries. The analysis illustrates that imposing a regulatory rule on an open access fishery, such as restricting harvest to a fixed proportion of the fishing stock, has important implications for valuing any linkage between coastal breeding and nursery habitat and a near-shore fishery.

The next section develops the model of a regulated open access fishery with coastal-habitat linkages and applies it to the case study of mangrove-dependent fisheries in Thailand. Section 3 presents the results of simulation of these fisheries and the valuation of habitat-fishery linkages. Section 4 discusses the implications of this analysis and its results for valuing coastal habitat-fishery linkages under regulated open access. Section 5 concludes with some observations concerning policy implications and future research.

2. Materials and Methods

2.1. A Regulated Open Access Fishery with Coastal-Habitat Linkages

The starting point for the valuation approach developed in this paper is the dynamic model of coastal-habitat-fishery linkages developed in Barbier [6]. The underlying assumption of the model is that a near-shore fishery depends on a coastal wetland habitat, such as salt marsh or mangroves, as a breeding habitat and nursery. Consequently, any change in the coastal wetland habitat is likely to affect the biological growth of the fishery, which is usually modeled through some influence on carrying capacity.

Defining X_t as the stock of fish measured in biomass units, any net change in growth of this stock over time can be represented as

$$X_{t+1} - X_t = F(X_t, S_t) - h(X_t, E_t), \quad \frac{\partial F}{\partial X_t^2} > 0, \quad \frac{\partial F}{\partial S_t} > 0 \tag{1}$$

Thus, any expansion in the fish stock occurs as a result of biological growth in the current period $F(X_t, S_t)$ net of any harvesting $h(X_t, E_t)$, which is the function of the stock as well as fishing effort E_t. The influence of the wetland habitat area S_t as a breeding ground and habitat on growth of the fish stock is assumed to be positive $\partial F/\partial S_t > 0$, as an increase in wetland area will mean more carrying capacity for the fishery and thus greater biological growth.

As the near-shore fishery is open access, effort in the next period will adjust in response to the profits made in the current period. Letting $p(h_t)$ represent landed fish price per unit harvested, w the unit cost of effort and $\phi > 0$ the adjustment coefficient, then total effort in the fishery changes according to

$$E_{t+1} - E_t = \phi[p(h_t)h(X_t, E_t) - wE_t], \quad \frac{\partial p}{\partial h_t} < 0 \tag{2}$$

Assume that biological growth is characterized by a logistic function $F(X_t, S_t) = rX_t(1 - X_t/K(S_t))$, and harvesting by a Schaefer production process $h_t = qX_tE_t$, where q is the catchability coefficient, r is the intrinsic growth rate and $K(S_t) = \alpha \ln S_t$ is the impact of coastal wetland area on carrying capacity K_t of the fishery. The market demand function for harvested fish is iso-elastic, i.e., $p(h_t) = kh_t^\eta$, $\eta = 1/\varepsilon < 0$. Substituting these expressions into Equations (1) and (2) yields

$$X_{t+1} - X_t = rX_t\left[1 - \frac{X_t}{\alpha \ln S_t}\right] - h_t, \quad h_t = qX_tE_t \tag{3}$$

$$E_{t+1} - E_t = \phi\left[kh_t^{\eta+1} - wE_t\right] \tag{4}$$

Following Homans and Wilen [13], it is assumed that an outside regulatory body imposes a simple quota rule to ensure that current harvest does not over-exploit the stock. Here, the regulatory rule is that harvest must always be a fixed proportion b of the current stock, i.e., $h_t = bX_t$. Thus, b represents the quota on current harvest-stock ratio, or the regulatory quota for short. Since the Schaefer production function dictates that $h_t = bX_t = qX_tE_t$, the implication of this rule is that effort will be fixed at $E = b/q$. Total effort in the fishery now depends on the regulator's decision on how large a proportion b of the stock can be safely fished. This can be stated in terms of the following proposition:

Proposition 1. If a regulatory rule is imposed that harvest is a fixed proportion b of the current stock, then total effort in the fishery is constant and determined by the size of the regulatory quota b, i.e., $E = b/q$.

Fixing effort in the fishery implies that Equation (4) becomes $k(bX_t)^{\eta+1} = wE = wb/q$. This implies that rents in the fishery will be totally dissipated. Intuitively, as total effort by all fishers is fixed, some will be able to fish enough to make profits, but others will not. The latter will leave the fishery to be replaced by those attracted by the profits, but overall effort will remain the same, and the process will repeat itself until eventually zero rents are earned throughout the fishery. The result is a unique value for fish biomass

$$X = ab^{-\frac{\eta}{\eta+1}}, \ a = \left(\frac{w}{kq}\right)^{\frac{1}{\eta+1}}, \ \frac{dX}{db} = -\frac{\eta}{\eta+1}ab^{-\frac{\eta}{\eta+1}-1} \tag{5}$$

Thus, fish stock is also constant in the regulated fishery. Note that the impact of a change in the regulatory quota on the stock will depend on the elasticity of demand for harvested fish ε. That is, if demand is relatively inelastic $-1 < \varepsilon < 0$ (implying $\eta < -1$), then $dX/db < 0$. However, for elastic demand $\varepsilon < -1$ ($-1 < \eta < 0$), then $dX/db > 0$. The following proposition follows:

Proposition 2. If the market demand for fish is relatively inelastic $-1 < \varepsilon < 0$, then the fish stock will decrease (increase) with a positive (negative) change in the regulatory quota b; if the market demand is relatively elastic $\varepsilon < -1$, the fish stock will increase (decrease) with a positive (negative) change in b.

As fish biomass is unchanging and governed by (5), then (3) becomes

$$r - b = \frac{rX}{\alpha \ln S_t} > 0 \tag{6}$$

Since the right-hand side of Equation (6) is positive, an important implication is:

Proposition 3. The regulatory quota b must always be less than the intrinsic growth rate r of the fish stock.

Equations (5) and (6) can be solved to determine the regulatory quota b that yields this equilibrium outcome for the fishery. Once b is known, it is possible to find X from Equation (5), E from Proposition 1, and then harvest h.

Valuing the impact of the change in coastal habitat area S_t can now be determined by examining how this change influences the equilibrium harvest outcome h and thus the consumer surplus for marketed fish. For example, if h^0 is the initial harvest in the fishery and h^1 is harvest after the change in coastal habitat area occurs, then the resulting change in consumer surplus CS will be

$$\Delta CS = \int_{h^0}^{h^1} p(h)dh - \left[p^1 h^1 - p^0 h^0\right] = -\frac{\eta\left[p^1 h^1 - p^0 h^0\right]}{\eta+1} \tag{7}$$

However, the value of this habitat–fishery linkage will depend on whether or not the regulatory quota b is adjusted in response to the change in S.

In the case where such an adjustment occurs, then the solution for b depends on the current size of the coastal habitat area, i.e., $b = b(S_t)$. From substituting Equations (5) in (6), one can find

$$r - b = \frac{rab^{-\frac{\eta}{\eta+1}}}{\alpha \ln S_t}, \frac{db}{dS} = \frac{(r-b)\alpha}{S_t\left[\alpha \ln S_t - \frac{\eta}{\eta+1} rab^{-\frac{\eta}{\eta+1}-1}\right]} \tag{8}$$

As Equation (8) is currently specified, the impact of a change in S_t on the regulatory quota db/dS_t is ambiguous. Once this effect is known, Propositions 1 and 2 can be invoked to determine the impacts on fishing effort E and stock X, respectively. The ensuing changes in harvest and consumer surplus follow.

In the case where the regulatory quota b does not change, fishing stock adjusts in response to the change in coastal habitat area, i.e., from Equation (6).

$$\frac{dX}{dS} = \frac{(r-b)\alpha}{rS_t} > 0 \tag{9}$$

As this impact is always positive, it leads to

Proposition 4. If the regulatory quota b does not adjust in response to a change in the coastal habitat area S_t, then fish stock will increase (decrease) in response to an increase (decrease) in S_t.

Because there is no adjustment to b, then fishing effort remains unchanged (Proposition 1). However, there is an impact on harvest, since $dh = bdX = \frac{b(r-b)\alpha}{rS} dS$. The result is a change in consumer surplus as indicated by Equation (7), which is the value attributed to the decline or increase in the coastal habitat supporting the fishery.

To summarize, the regulatory rule $h_t = bX_t$ imposed on the open access fishery with coastal habitat linkages leads to a bioeconomic equilibrium that is fully recursive. Valuing any changes in the coastal habitat supporting the fishery will depend on whether or not the regulatory quota b is adjusted in response to these changes. If the regulatory quota adjusts for any change in S_t, the result will be changes in both fishing effort and stock. Although rents are still fully dissipated, the resulting change in harvest will impact consumer surplus. If b does not adjust, then effort is unchanged but the stock of fish will respond to any change to coastal habitat area. Both harvest and consumer surplus are again impacted. However, these differing impacts are sufficiently important that they can affect significantly the value attributed to the coastal-habitat-fishery linkage.

2.2. Case Study: Mangrove-Dependent Fisheries, Thailand

The above approach for valuing coastal habitat support for a regulated fishery is illustrated through application to the mangrove-dependent demersal and shellfish fisheries in Thailand, based on data from Barbier [6].

Up to 38,000 households in around 2500 coastal communities engage in small-scale fishing activities, which are largely open access [6]. These communities are located along the Southern Gulf of Thailand and Andaman Sea (Indian Ocean) coasts. Gill nets and both motorized and non-motorized small boats are the most common form of fishing gear used by artisanal fishers. Although a license fee and permit are required for fishing in coastal waters, officials do not strictly enforce the law and users do not pay. Currently, there is no legislation for supporting community-based fishery management, and regulation of the fisheries is negligible.

Based on data from Barbier [15] that identifies which of Thailand's artisanal demersal and shellfish fisheries depend on mangroves for breeding and nurseries, Barbier [6] employs pooled time-series and cross-sectional regressions that yield the key biological parameters (r, α), economic parameters (k, w, q) for the two fisheries. This allows determination of the key relationships in Equations (5)–(9) necessary for valuing coastal-habitat linkages in a regulated open access fishery. In addition, evidence from domestic fish markets in Thailand suggest that the demand for fish is fairly inelastic, and an elasticity of −0.5 is assumed for the iso-elastic market demand function. These key parameter estimates for Thailand's mangrove-dependent demersal and shellfish fisheries are summarized in Table 1. In addition, the table indicates the estimated area of mangroves along the Gulf of Thailand and Andaman Sea supporting these fisheries.

Table 1. Key parameters for mangrove-dependent fisheries, Thailand.

Parameter	Demersal Fishery	Shellfish Fishery
r	0.4896	0.2997
α	12,817,069	24,663,448
k	1.6×10^{13}	6.2×10^{14}
w	3415	113,155
q	0.0002	0.00006
ε	−0.5	−0.5
η	−2	−2
S_t	1672 km^2 (1996)	1672 km^2 (1996)

Source: Barbier [6].

The parameter estimates in Table 1 are used in the model developed here to estimate the value of mangrove–fishery linkages for an open access fishery that is regulated according to the rule $h_t = bX_t$. The first step is solving conditions in Equations (5) and (6) to find the corresponding regulatory quota b, and then X, E, and h for each fishery. The second step is determining the changes in consumer surplus resulting from annual average changes in mangrove deforestation, depending on whether the regulatory quota adjusts or not.

In this valuation exercise, two different calculations of annual mangrove deforestation rates are used, based on high and low estimates for Thailand over 1996 to 2004 [6]. The low estimate by the Royal Thai Forestry department is 3.44 km^2; the high value by the UN Food and Agricultural Organization is 18.0 km^2.

3. Results

3.1. Simulation of the Regulated Mangrove-Dependent Fisheries

Table 2 depicts the outcomes for the regulatory quota b, fish stock X, fishing effort E, and harvest h for the mangrove-dependent demersal and fisheries, respectively, derived through applying the parameter estimates from Table 1 to the model and assuming that mangrove area remains at 1672 km^2.

Table 2. Simulation results for mangrove-dependent fisheries, Thailand.

Variable	Demersal Fishery	Shellfish Fishery
b	0.1146	0.0462
X	72,858,304 kg	154,821,198 kg
E	561×10^3 hours	766×10^3 hours
h	8,351,529 kg	7,156,579 kg
S_t	1672 km^2 (1996)	1672 km^2 (1996)

The model simulation applied to the demersal fishery suggests a regulatory quota of about 11.5% of fish stock. For the shellfish fishery, b is much lower, around 4.6%. However, because the stock of shellfish is greater than for the demersal fishery, the resulting equilibrium harvests are approximately the same. The regulated harvest $h = bX$ is around 8400 metric tons for the demersal fishery and 7200 metric tons for shellfish.

3.2. Valuation of Habitat-Fishery Linkages

Table 3 depicts the resulting welfare impacts of mangrove deforestation on the demersal and shellfish fisheries, using the low and high estimates of annual deforestation. In addition, the table shows how these welfare impacts vary, depending on whether the regulatory quota adjusts or not in response to mangrove loss.

Table 3. Valuation of mangrove-fishery linkages under regulated open access, Thailand.

Annual Mangrove Loss	Demersal Fishery		Shellfish Fishery	
	3.44 km^2	18.0 km^2	3.44 km^2	18.0 km^2
b constant:				
Welfare loss ($/km^2)	$309	$309	$13,960	$13,960
Welfare loss ($/year)	$1062	$5564	$48,031	$251,621
b adjusts:				
Welfare loss ($/km^2)	$14	$14	$103	$103
Welfare loss ($/year)	$47	$245	$354	$1854

The results show that the valuation estimates of the loss in mangrove habitat-fishery support service vary considerably for the two fisheries, and depending on whether or not the regulatory quota *b* is adjusted in response to mangrove loss. As expected, the annual welfare loss is much larger for the high versus low estimate of mangrove deforestation.

The loss in value of the mangrove–fishery linkage is much greater for the shellfish fishery as opposed to the demersal fishery. When *b* is constant, the loss is nearly $14,000 per km^2 of mangrove decline for shellfish compared to $309 per km^2 for the demersal fishery. When *b* can adjust, the loss is $103 per km^2 for shellfish and only $14 for the demersal fishery. This suggests that the mangrove-habitat–fishery linkage is a more valuable service for the shellfish than the demersal fishery.

As these estimates suggest, the welfare loss for both fisheries is always higher when the regulatory quota is fixed as opposed to when it can be adjusted. When the quota is constant, effort in the fishery cannot adjust (Proposition 1), and as a result, there is a much larger fall in harvest. The corresponding loss in consumer surplus is therefore much greater (see Equation (7)). Based on the parameter estimates in Table 1, Equation (7) becomes $\Delta CS = -2k\left[(h')^{-1} - h^{-1}\right]$, $h' = h + dh$.

Finally, at low rates of mangrove loss, the welfare decline in the habitat–fishery linkages is $1062 per year for the demersal fishery and $48,031 per year for the shellfish fishery. However, for the high deforestation estimate, the welfare decline is $5564 per year for the demersal fishery and $251,621 annually for shellfish.

4. Discussion

This analysis has shown that imposing a regulatory rule on an open access fishery, such as restricting harvest to a fixed proportion of the fishing stock, has important implications for valuing any linkage between coastal breeding and nursery habitat and a near-shore fishery.

First, total effort in the fishery is constant and determined by the size of the regulatory quota *b* chosen (Proposition 1), and in turn, *b* must always be less than the intrinsic growth rate *r* of the fishery (Proposition 3). These two outcomes are linked. As shown in the model, the effect of restricting harvest to a fixed proportion of stock is that the fishery will converge to a bioeconomic equilibrium with constant effort and fish biomass. As *r* represents how rapidly the fish population grows to reach carrying capacity, it must serve as an upper bound on how a large of a proportion of the stock can be harvested through the regulatory quota *b*. This condition holds for the simulation applied to mangrove-dependent fisheries in Thailand (see Tables 1 and 2). For example, in the demersal fisheries, the intrinsic growth rate is 0.49 and the corresponding equilibrium regulatory quota is estimated to be 0.11. In the shellfish fisheries, *r* is 0.30 and the corresponding *b* is 0.05.

Second, in the regulated open access fishery the elasticity of demand also impacts the magnitude of the welfare effects associated with a change in a supporting coastal habitat. However, in the regulated fishery, the values associated with any changes in the coastal habitat supporting the fishery will depend even more on whether or not the regulatory quota b is adjusted in response to these changes. As the simulation of the regulated mangrove-dependent fisheries shows, the welfare loss imputed to coastal habitat decline is always much higher when the regulatory quota is fixed opposed to when it can be adjusted.

Intuitively, this makes sense. When b cannot be adjusted, effort remains fixed and so the loss of habitat area has a direct negative impact on the fish stock (see Proposition 4). Harvest must therefore fall and so must consumer surplus. However, when b adjusts to accommodate for the fall in coastal habitat area, then effort must adjust also. In the mangrove-dependent fishery simulation of this paper, the correct adjustment is actually to increase b in response to an increase in S. This allows effort to increase to offset for a decline in X that must occur with the increase in the quota (see Proposition 2). The overall net effect is still a decline in harvest and thus consumer surplus, but not as much if the increase in E did not offset some of impact on the decline in harvest from a smaller stock.

Of course, in practice it is likely to be difficult for any regulatory body to adjust its imposed fishing quota accurately in response to a change in coastal breeding and nursery habitat. Even for the simple regulated fishery model developed here, determining how to modify b for a change in S requires knowledge of a considerable number of key bioeconomic and habitat parameters for each fishery (see Table 1), correct estimation of key initial variables (see Table 2), and accurate determination of the relationship $b = b(S_t)$ such as in Equation (8).

Third, as the Thailand mangrove-dependent fishery simulation illustrates, the value of the habitat support service will vary from fishery to fishery. As the simulation shows, the value of mangroves as a breeding and nursery ground is significantly greater for shellfish compared to the demersal fishery. This suggests that, regardless of whether the fishery is regulated or open access, the type of fish populations supported by mangroves and other estuarine and coastal ecosystems will influence considerably the value of coastal-habitat-fishery linkages.

Finally, although the discussion so far has focused on the implications for managing and regulating the fishery, this analysis also demonstrates the need to manage critical estuarine and coastal ecosystems that support near-shore fisheries. As Table 2 indicates, the loss in mangroves has a significant impact on Thailand's demersal and shellfish fisheries that are essential to the livelihoods of coastal communities. Thailand's mangroves are also valuable for protection against coastal storms and for a variety of products collected by local people from the forests [6]. Given the extensive benefits of estuarine and coastal ecosystems, protection and restoration of these critical habitats are important not only for maintaining habitat-fishery linkages but also for many other important ecosystem services emanating from the coastal zone [1–3].

5. Conclusions

The purpose of this paper has been to explore how regulating an open access fishery may affect the value attributed to an estuarine and coastal habitat that serves as a breeding and nursery ground for the fishery. Although there is a growing number of studies that model and value this habitat-fishery linkage; to date, none have done so under conditions of regulated open access.

Models of open access fisheries indicate that the magnitude of the elasticity of demand is a significant influence on the value of any coastal-habitat linkage. However, both the model developed here for regulated open access and its application to mangrove-dependent fisheries in Thailand indicate that this value is influenced even more by whether or not the regulatory quota can adjust in response to the changes in a supporting coastal habitat. This outcome is analogous to what happens in regulated fishery when one input (capital) is restricted by a regulation, as opposed to all inputs are able to vary freely [12]. Similarly, such "target flexibility" is critical to diminishing the economic losses from environmental shocks that impact both costs and growth [10]. As more and more fisheries are regulated in some manner, usually through simple rules that tie harvest to 'safe' stock levels, the results of our model have important implications for both policy and future research directions.

First, even under regulated open access, the value of the coastal habitat service will vary from fishery to fishery. For example, as the Thailand case study demonstrates, the value of mangroves as a breeding and nursery ground is significantly greater for shellfish compared to the demersal fishery. Identifying which fish populations are most impacted by losses of coastal habitats remains an important policy goal, regardless of whether fisheries are open access or regulated. Similarly, identifying which of these habitats are threatened by rapid and large-scale coastal developments should also be a policy priority.

Second, the regulatory rule must be chosen carefully, and in reference to the biological and economic conditions of the fishery. In the model of this paper, the intrinsic growth rate of the fishery serves as an upper bound on the regulatory quota imposed. However, for more sophisticated models of coastal-fishery-habitat linkages with complex regulatory rules, other restrictions on the choice of the regulatory rule may apply. These need to be investigated further, especially for a regulatory fishery that is supported by estuarine and coastal ecosystems that serve as breeding and nursery grounds.

Third, although it may be difficult to adjust any regulatory restriction accurately in response to a change in coastal breeding and nursery habitat, such an adjustment could significantly mitigate the losses associated with this change. A regulatory body must consider how best to monitor its policies, to ensure that regulatory rules and measures are sufficiently flexible to accommodate the economic impacts of habitat-fishery linkages. In some instances, the body might opt for taxes and other instruments that do not severely restrict effort or harvest, as opposed to harvest or effort quotas. Choosing the correct instrument to regulate fisheries that benefit from coastal habitat-fishery linkages, and determining whether the instrument is sufficiently flexible in response to losses in coastal habitats, are important areas for future research.

Fourth, appropriate regulation of fisheries in response to loss of critical coastal breeding and nursery habitats is important, but preventing the loss of these habitats must also be recognized as an essential strategy for maintaining the long-term viability of near-shore fisheries. In recent years, linking such management approaches has become integral to ecosystem-based management of fisheries, which is starting to have impacts on fishery policy in the United States and other countries [16,17].

Finally, this paper does not address how environmental uncertainty over habitat–fishery linkages may affect its value to a regulated open access fishery. The type of uncertainty and its effects can determine both the choice of regulatory instrument and whether or not it needs to be sufficiently flexible in light of the environmental risks encountered [8]. Also important is the uncertainty of the links between coastal habitats, near-shore fisheries, and more distant marine habitats, or how such habitat-fishery linkages vary across entire "seascapes" from shoreline to sea [2,18]. Addressing these uncertainties could be critical to future research in valuing coastal habitat–fishery linkages under regulated open access.

Funding: This research received no external funding.

Acknowledgments: I am grateful to the support and encouragement of Wally Milon and the helpful comments provided by two anonymous reviewers.

Conflicts of Interest: The author declares no conflict of interest.

References

1. Barbier, E.B. Progress and challenges in valuing coastal and marine ecosystem services. *Rev. Environ. Econ. Pol.* **2012**, *6*, 1–19. [CrossRef]
2. Drakou, E.G.; Pendleton, L.; Effron, M.; Carter Ingram, J.; Teneva, L. When ecosystems and their services are not co-located: Oceans and costs. *ICES J. Mar. Sci.* **2017**, *74*, 1531–1539. [CrossRef]
3. Turner, R.K.; Schaafsma, M. (Eds.) *Coastal Zones Ecosystem Services: From Science to Values and Decision Making*; Studies in Ecological Economics; Springer: New York, NY, USA, 2015; Volume 9.

4. Foley, N.S.; Armstrong, C.W.; Kahul, V.; Mikkelsen, E.; Reithe, S. A review of bioeconomic modelling of habitat-fishery interactions. *Int. J. Ecol.* **2012**, 861635. [CrossRef]

5. Freeman, A.M., III. Valuing environmental resources under alternative management regimes. *Ecol. Econ.* **1991**, *3*, 247–256. [CrossRef]

6. Barbier, E.B. Valuing ecosystem services as productive inputs. *Econ. Pol.* **2007**, *22*, 177–229. [CrossRef]

7. Barbier, E.B.; Strand, I.; Sathirithai, S. Do open access conditions affect the valuation of an externality? Estimating the welfare effects of mangrove-fishery linkages in Thailand. *Environ. Res. Econ.* **2002**, 343–367. [CrossRef]

8. Smith, M.D. Generating value in habitat-dependent fisheries. The importance of fishery management institutions. *Land Econ.* **2007**, *83*, 59–73. [CrossRef]

9. Fischer, C.; Guttormsen, A.G.; Smith, M.D. Disease risk and market structure in salmon aquaculture. *Water Econ. Pol.* **2017**, *2*, 1650015. [CrossRef]

10. Kennedy, C.J.; Barbier, E.B. Renewable resource harvesting under correlated biological and economic uncertainties: Implications for optimal and second-best management. *Environ. Res. Econ.* **2015**, *60*, 371–393. [CrossRef]

11. Reimer, M.N.; Wilen, J.E. Regulated open access and regulated restricted access fisheries. *Encycl. Energy Nat. Resour. Environ. Econ.* **2013**, *2*, 215–223.

12. Deacon, R.E.; Finnoff, D.; Tschirhart, J.T. Restricted capacity and rent dissipation in a regulated open access fishery. *Res. Energy Econ.* **2011**, *33*, 366–380. [CrossRef]

13. Homans, F.R.; Wilen, J.E. A model of regulated open access resource use. *J. Env. Econ. Man.* **1997**, *32*, 1–21. [CrossRef]

14. Homans, F.R.; Wilen, J.E. Markets and rent dissipation in regulated open access fisheries. *J. Env. Econ. Man.* **2005**, *49*, 381–404. [CrossRef]

15. Barbier, E.B. Habitat-fishery linkages and mangrove loss in Thailand. *Contemp. Econ. Pol.* **2003**, *21*, 59–77. [CrossRef]

16. McLeod, K.L.; Leslie, H. (Eds.) *Ecosystem-Based Management for the Oceans*; Island Press: Washington, DC, USA, 2009.

17. Dolan, T.E.; Patrick, W.S.; Link, J.S. Delineating the continuum of marine ecosystem-based management: A U.S. fisheries reference point perspective. *ICES J. Mar. Sci.* **2016**, *73*, 1042–1050. [CrossRef]

18. Barbier, E.B.; Lee, K.D. Economics of the marine seascape. *Int. Rev. Environ. Res. Econ.* **2014**, *7*, 35–65. [CrossRef]

water

MDPI

Article

Valuing Provision Scenarios of Coastal Ecosystem Services: The Case of Boat Ramp Closures Due to Harmful Algae Blooms in Florida [†]

Sergio Alvarez [1],*, Frank Lupi [2], Daniel Solís [3] and Michael Thomas [3]

[1] Rosen College of Hospitality Management, University of Central Florida, Orlando, FL 32819, USA
[2] Department of Agricultural, Food, and Resource Economics, Michigan State University, East Lansing, MI 48824, USA; lupi@msu.edu
[3] College of Agriculture and Food Sciences, Florida A&M University, Tallahassee, FL 32307, USA; daniel.solis@famu.edu (D.S.); michael.thomas@famu.edu (M.T.)
* Correspondence: sergio.alvarez@ucf.edu; Tel.: +1-407-903-8001
† Authorship ordered alphabetically, no senior authorship implied.

Received: 13 May 2019; Accepted: 11 June 2019; Published: 14 June 2019

Abstract: Ecosystem service flows may change or disappear temporarily or permanently as a result of environmental changes or ecological disturbances. In coastal areas, ecological disturbances caused by toxin-producing harmful algae blooms can impact flows of ecosystem services, particularly provisioning (e.g., seafood harvesting) and cultural services (e.g., recreation). This study uses a random utility model of recreational boating choices to simulate changes in the value of cultural ecosystem services provided by recreation in coastal ecosystems resulting from prolonged ecological disturbances caused by harmful algae blooms. The empirical application relies on observed trips to 35 alternative boat access ramps in Lee County, an important marine access destination in southwest Florida. Results indicate that reduced boating access from harmful algae blooms may have resulted in losses of $3 million for the 2018 blooms, which lasted from the end of June to the end of September.

Keywords: harmful algae blooms; cyanobacteria; recreational boating; ecosystem services; random utility model; economic analysis

---

1. Introduction

Coastal ecosystems provide a diversity of services that contribute to social well-being. While human use and enjoyment of some of these services are captured (and measurable) by market transactions, most uses of these vital ecosystem services are not. Among these non-market ecosystem services, perhaps the most readily measurable is recreational use of waterways, particularly services related to recreational boating. Although recreational boating does not account for the total value of coastal ecosystems and the services they provide, recreational boating in Florida (FL) is an important cultural service[A] and a key component of the value of coastal ecosystem services. In 2017 there were close to 12 million registered recreational boats in the United States (US), and nearly 1 million of these were in FL [1]. These boaters enjoy the cultural services provided by clean waterways and healthy coastal ecosystems. Understanding the monetary value of these services can help coastal managers and policy-makers in their decision-making processes.

For boaters, the boat ramp infrastructure provides access to cultural ecosystem services produced in coastal waters. These ecosystem services produce economic benefits to the boaters in the form of increased well-being and satisfaction from boating, and these benefits are above and beyond the direct costs of boating (e.g., transportation costs of pulling boats to the ramp, fuel for the boats, fees, etc.). Economists refer to such benefits as consumer surplus, and this surplus comprises one side in the benefit-cost equation in situations where the benefits of boating-related coastal ecosystems services may be weighed against the costs of protecting, conserving, or restoring the ecosystems that produce them.

In this study, we develop a model that estimates changes in the consumer surplus from boating to inform a policy-relevant scenario related to ecological disturbances from harmful algae blooms (HABs). Specifically, the economic models developed here are models of the demand for access to boating sites and are suitable for valuing access as well as the characteristics of boating sites. We use random utility models (RUMs) informed by data on individual trips to explain boaters' site choices and to relate these choices to the costs of access and characteristics of alternative boating sites [2]. Boaters' choices reveal their relative preferences for site characteristics and travel costs, i.e., the boaters' willingness to trade costs (or money) for site characteristics or specific destinations. Through this linkage, RUMs can value changes in access to specific sites or individual characteristics, and they can be used to measure the welfare implications of changes in the provision and quality of ecosystem services. Similar models have been successfully employed to measure these changes for a wide variety of coastal ecosystem services [3–7]. As mentioned by [8], the model contributes to the relatively thin existing literature valuing recreational boating.

In this article, we use data and estimates from a baseline study and data collection commissioned by the FL Fish and Wildlife Conservation Commission [9]. A RUM model is used to simulate losses in consumer surplus as a result of recent and ongoing harmful algae blooms (HABs) of cyanobacteria from the genus *Microcystis* and *Anabaena*, which besides resulting in unsightly green tinted water and foul smells also produce toxins that lead to fish kills and adverse impacts to human health [10,11]. To accomplish this, we develop scenarios of ramp closures as a result of these HABs, and we examine the impact of duration of the algae blooms on the loss of consumer surplus from boating. This study contributes to thin literature on economic valuation of HABs on ecosystem services [12]. In addition to presenting the boating RUM model results, the paper illustrates an approach for using economic valuation studies to inform policy-relevant scenarios of environmental degradation that arise after the initial study is commissioned.

The rest of this article is organized as follows. Section 2 introduces the problem of HABs in Florida, specifically in the study region, and discusses how this problem impacts the ecosystem services enjoyed by recreational boaters. Section 3 outlines the methodology, and Section 4 presents the data used to develop the RUM. Section 5 provides an overview of the results of the empirical application as well as the evaluation of closure scenarios to provide a tangible application for our framework. The paper ends with a discussion and conclusion in Sections 6 and 7.

2. Harmful Algae Blooms and Their Impact on Recreational Boating

Coastal and waterfront communities depend on clean water as the foundation of a healthy and growing economy. Visitors from around the world are drawn to clean and pristine beaches, lakes, and other waterways, and local communities benefit from the economic activity that occurs when visitors spend their money in hotels, restaurants, shops, and other services. Similarly, residents of waterfront communities regularly use beaches and other waterways for recreation and community-building. In short, clean water is the thread that ties waterfront communities together, drives their economies, and provides a quality of life for residents and a positive experience for visitors.

However, issues related to eutrophication and HABs are having negative consequences on social well-being in many parts of the world [13–18]. In numerous FL waterfront communities, widespread HABs are severely impacting water quality and are having a deleterious impact on local economies and human health. HABs occur when colonies of photosynthetic microorganisms that live in fresh or saltwater grow out of control and produce toxins that can have harmful effects on people or wildlife. In recent years, the state of Florida has experienced several HABs, most notably outbreaks of red tide (*Karenia brevis*) and cyanobacteria (*Microcystis spp.*, and *Anabaena spp.*).

HABs have become more frequent in recent years as a result of warmer temperatures and nutrient pollution, and threaten to become a chronic issue for many communities in FL. Not surprisingly, the spread of these blooms has also affected the tourism industry in the state, which has a substantial impact on FL's economy. Namely, the 118.5 million people that visit FL each year provide an infusion of $111.7 billion to the state's economy [19].

The 2018 HABs provide a worrisome picture of what this problem may look like in the future. During the month of June, a significant cyanobacteria bloom emerged in Lake Okeechobee and was transported to both the southeast and southwest coasts of Florida via the St. Lucie and Caloosahatchee rivers respectively [20]. However, in southwest Florida, the situation was compounded by an unprecedented bloom of red tide that moved into the region's beaches and waterways from the Gulf of Mexico. Consequently, some communities in southwest FL experienced two different types of HABs: Cyanobacteria originating inland in Lake Okeechobee, and red tide originating offshore in the Gulf of Mexico.

These unprecedented HABs on both coasts have significantly impacted local economies, public health, and the environment. HABs are likely to have long-lasting impacts on residents' well-being [10,11] and may also damage FL's brand as a world-class tourism destination. Nationwide, eutrophication, and HABs occurring in freshwater—mostly outbreaks of cyanobacteria—cost an estimated $2.2 billion per year [21]. However, there is a notable absence of research quantifying the impact of these blooms on local economies and visitor's choices, and most of the related research that has been done in FL has focused on the impacts of red tide [22].

The cyanobacteria prevalent in FL are naturally occurring in freshwater lakes, and they are generally present in very low concentrations throughout the year. However, high nutrient loads of phosphorus and nitrogen provide the fuel that allows these cyanobacteria to reproduce into uncontrolled noxious blooms [23]. In FL, a large portion of the state's urban and rural runoff flows into the Kissimmee River Basin, where water is conveyed from densely populated central FL southward toward Lake Okeechobee. Historically, water in Lake Okeechobee would continue flowing south through the Everglades, but this historic flow has been modified through a system of canals and the construction of an earthen dike to prevent flooding in response to a 1928 hurricane that resulted in more than 2500 deaths [24]. Today, rather than flow south toward FL Bay through the Everglades, water from Lake Okeechobee flows east through the St. Lucie River into the St. Lucie Estuary, and west through the Caloosahatchee River into the Gulf of Mexico (Figure 1). Lake discharges are managed by the US Army Corps of Engineers through a system of locks and gates [25].

Cyanobacterial HABs in Lee County are fairly predictable events that originate in the nutrient-rich waters of Lake Okeechobee and are transported to the Caloosahatchee River via Lake discharges through managed locks and gates. As the cyanobacterial blooms are transported into saltwater the bacteria's cell membranes are compromised, the cells die, and any cyanotoxins in the cell are released into the water [25,26]. The cyanobacteria blooms, therefore, dissipate as they reach areas with high water salinity in the Gulf of Mexico. In recent years, cyanobacterial HABs have been observed throughout the Caloosahatchee River, but the HABs have tended to dissipate as they reach the higher salinity areas where the mouth of the river reaches the Gulf of Mexico (Figure 2).

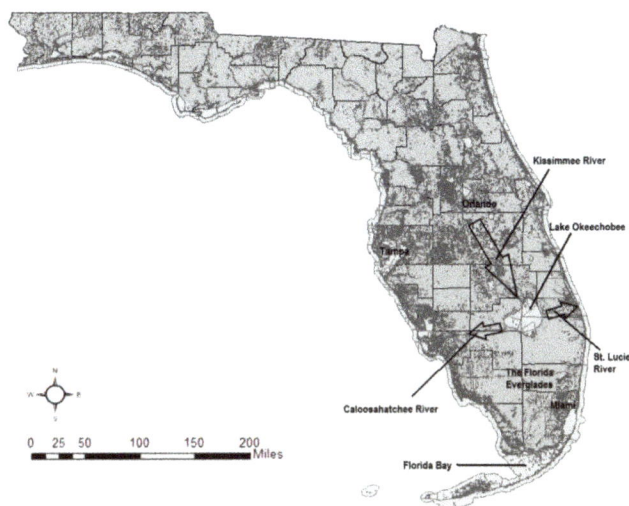

Figure 1. Map of Florida showing current water flow into and out of Lake Okeechobee. The locations of the densely populated cities of Orlando, Tampa, and Miami are noted, as well as the locations of the Kissimmee River, Lake Okeechobee, the Caloosahatchee River, the St. Lucie River, the Florida Everglades, and Florida Bay. Arrows show the current direction of water flow impacting the study region, which begins in the Orlando area in Central Florida and flows south into Lake Okeechobee. Prior to construction of Florida's system of water control structures and canals, water would flow south from Lake Okeechobee into the Florida Everglades, and continue flowing in a southward direction into Florida Bay. After construction of the water control system, water instead flows west to the Caloosahatchee River, and east to the St. Lucie River.

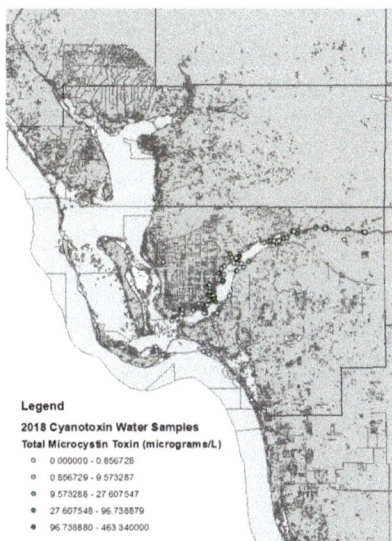

Figure 2. Water sampling locations and concentrations of cyanotoxins in Lee County throughout the 2018 cyanobacterial HABs. High concentrations of microcystin were observed in several locations along the Caloosahatchee River, but only very low or no concentrations were observed beyond the mouth of the river where salinity is higher [27].

3. A Random Utility Model to Value Site Closures Due to HABs

When estimating a model of demand for public goods such as boat ramps and beaches, the RUM approach is particularly well suited when there are many identifiable substitutes from which to choose [28,29]. This method has been widely used in the region. In the mid-1990s, the FL Department of Environmental Protection successfully used a RUM to estimate the recreational value that was lost to beach visitors following the 1993 Tampa Bay oil spill [30]. References [3–6,31–33] have applied RUMs to estimate changes in welfare resulting from perturbations in recreational fishing and boating. RUMs have also been used to evaluate the welfare lost to boaters from policies designed to protect the West Indian manatee in Lee County (restricted boating speeds and waterway access) and were later extended to Brevard County in 2003 [7,34].

In our application, it is assumed that a boater will choose a combination of a launch ramp and on-the-water destination among many possible alternatives on each choice occasion. The factors that affect choice include the cost of traveling to the ramp, the cost of boating to the desired on-the-water destination, the perceived quality of the location as a recreation site, and other characteristics of the ramp and site. We can model the individual's conditional indirect utility from visiting site j as a linear function of trip costs and site characteristics given by tc_j and q_j:

$$v_j = \beta_{tc} tc_j + \beta_q q_j + \varepsilon_j, \tag{1}$$

where tc_j is the cost of traveling to the site j, q_j is a vector of the site j characteristics, ε_j is a random error term accounting for factors that remain unobservable, and the βs are parameters to be estimated. The absolute value of the travel cost parameter β_{tc} is hypothesized to be negative and serves as a measure of the marginal utility of income. The elements of vector β_q are the marginal utilities of site characteristics and are expected to be positive if the characteristics are desirable and negative if undesirable. Following RUM theory, an individual is assumed to select the site with the highest utility. Thus, the probability of an individual choosing site i is given by:

$$\Pr\left(\beta_{tc} tc_i + \beta_q q_i + \varepsilon_i > \beta_{tc} tc_j + \beta_q q_j + \varepsilon_j\right) \text{ for all } i \neq j. \tag{2}$$

Assuming the random errors are independently and identically distributed type I extreme value, Equation (2) can be estimated by a conditional logit model. In our case, we expect that the errors associated with on-the-water destinations are more correlated with one another than they are with error terms associated with the boat ramps, so we adopt a nested logit model in which the water destination sites are nested below ramp sites. Although the decision of ramp and on-the-water site is assumed to be made simultaneously, this two-level nesting structure can be modeled as an individual choosing a ramp and then choosing the water site conditional upon the selected ramp.

Let k represent Lee County ramps and j represent on-the-water sites, which are nested by each ramp j. A combination of on-the-water destination reached from a particular ramp is represented by a combination of (j, k). The equations can be rewritten to reflect these nests as:

$$v_{jk} = \beta_{tc} tc_{jk} + \beta_q q_{jk} + \varepsilon_{jk} \tag{3}$$

$$\Pr\left(\beta_{tc} tc_{il} + \beta_q q_{il} + \varepsilon_{il} > \beta_{tc} tc_{jk} + \beta_q q_{jk} + \varepsilon_{jk}\right) \text{ for all } i \neq j \text{ and } l \neq k \tag{4}$$

Let $\Pr(j,k)$ be the probability of choosing site (j, k) from among all feasible combinations, that is the probability that indirect utility from site (j, k) exceeds the indirect utility from any other site. Assuming error terms ε_{jk} are distributed as generalized extreme value, then following [35] the probability of choosing site (j, k) is:

$$\Pr(j,k) = \frac{\exp\left(v_{jk}/\theta\right)\left[\sum_{j=1}^{J_k} \exp\left(v_{jk}/\theta\right)\right]^{\theta-1}}{\sum_{k=1}^{K}\left[\sum_{j=1}^{J_k} \exp\left(v_{jk}/\theta\right)\right]^{\theta}} \tag{5}$$

where θ is the nested logit distributional parameter to be estimated. To clarify our estimation approach, consider $\Pr(j, k)$ as the product of the conditional probability of choosing site j, given ramp k, $\Pr(j|k)$, times the marginal probability of choosing ramp k, $\Pr(k)$. That is,

$$\Pr(j, k) = \Pr(j|k)\Pr(k) = \frac{\exp(v_{jk}/\theta)}{\sum_{j=1}^{J_k} \exp(v_{jk}/\theta)} \times \frac{\left[\sum_{k=1}^{J_k} \exp(v_{jk}/\theta)\right]^{\theta}}{\sum_{k=1}^{K}\left[\sum_{j=1}^{J_k} \exp(v_{jk}/\theta)\right]^{\theta}}, \tag{6}$$

where $\Pr(k)$ and $\Pr(j|k)$ are given by:

$$\Pr(j|k) = \frac{\exp(v_{jk}/\theta)}{\sum_{j=1}^{J_k} \exp(v_{jk}/\theta)} \quad \text{and} \tag{7}$$

$$\Pr(k) = \frac{\left[\sum_{k=1}^{J_k} \exp(v_{jk}/\theta)\right]^{\theta}}{\sum_{k=1}^{K}\left[\sum_{j=1}^{J_k} \exp(v_{jk}/\theta)\right]^{\theta}} \tag{8}$$

another expression for $\Pr(k)$ is:

$$\Pr(k) = \frac{\exp(\theta IV_k)}{\sum_{k=1}^{K} \exp(\theta IV_k)}, \tag{9}$$

where $IV_k = \ln(\sum_{j=1}^{J_k} \exp(v_{jk}/\theta))$ is known as the inclusive value for ramp k and θ is the inclusive value parameter. Note too that if the utility function contains characteristics that do not vary across water sites but do vary across ramps, which is a realistic assumption, we can re-write Equation (9) as:

$$P_{ijk} = \frac{\exp(\beta Z_k + \theta IV_{j|k})}{\sum_{k=1}^{K} \exp(\beta Z_k + \theta IV_{j|k})}, \tag{10}$$

where the Z_k represent characteristics of the ramps.

In this notation, both choice probabilities take the conditional logit form. A consistent estimation strategy for the nested logit is to estimate two conditional logits linked by the lower level inclusive value index. We present the sequentially estimated model below with the first part corresponding to on-the-water site choices conditional on a ramp chosen, and the second part corresponding to the ramp choices as a function of the inclusive value of the on-the-water sites available from each ramp.

The resulting estimated model can be used for policy analysis, as the measure of welfare change estimated (benefits or damages) follows the earlier work of [36]. The welfare change resulting from the removal of sites from the choice set due to HABs can then be calculated as the consumer surplus, which, if we abstract from the nesting for the sake of exposition, is:

$$CS = -\frac{1}{\beta_{tc}}\left[\ln\sum_{k=1}^{K} e^{V_{noHAB}} - \ln\sum_{k=1}^{K} e^{V_{HAB}}\right] \times T \tag{11}$$

where V_{noHAB} are the indirect utilities derived by boaters from the full choice set with K ramps when there are no cyanobacterial HABs, V_{HAB} are the indirect utilities derived by boaters when cyanobacterial HABs are present (which can indicate diminished utility or ramp closure), β_{tc} is the parameter for travel cost that represents the marginal utility of money, and T is the estimated number of choice occasions impacted by HABs.

4. Data

The first step in estimating the choice model is to define those ramps that are available to the boating public (see Supplementary Materials). In September of 2006, a field team led by the FL Fish and Wildlife Conservation Commission conducted a pilot study in Lee County that surveyed all ramp in the area. Of the 97 Lee County inventoried ramps, 55 ramps were not available for public use for a variety of reasons including temporary closures, private or gated facilities, and government ramps open only for official use. Included in the remaining 42 ramps are the obvious stand-alone public ramps and public access marinas with launch lanes. Ramps that are closed to public access were excluded from the analysis.

Next, the location of ramps in relation to each other was considered. When choosing an access point, boaters likely consider ramps in close proximity to one another as members of a larger group or aggregate. For example, if the parking lot of one ramp is full the boater could easily move along to the nearby neighboring ramp with no significant increase in travel time or cost, and still be able to reach the desired on-the-water site. Therefore, nearby ramps were aggregated as a single ramp to capture this choice behavior. Specifically, ramps within 1.5 road miles of each other were grouped and considered single aggregated ramps. For Lee County, twelve ramps were aggregated into five groups leaving a total of 34 individual ramp choices (Figure 3).

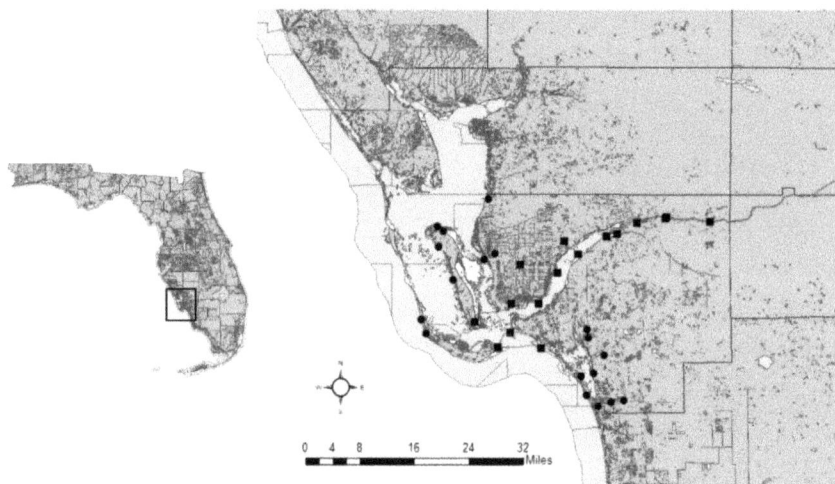

Figure 3. Map of Florida and location of boat ramps in Lee County. Ramps represented by squares experience cyanobacteria harmful algae blooms originating from Lake Okeechobee discharges. Boat ramps represented by circles are assumed not to be impacted by cyanobacteria algae blooms.

With the ramps selected, the next step in preparing the data involved identifying on-the-water destination sites. Florida Fish and Wildlife Conservation Commission (FWC) constructed a statewide GIS grid overlay comprised of 73,485 one-mile-square cells. Each grid cell contained at least 30 variables representing cell attributes including the presence or absence of salt and/or fresh water, natural and/or artificial reefs, seagrass, navigational aids, manatee protection status and marine protection/conservation status. Information also included bathymetry data and lake acreage, among other variables. For Lee County, the one-mile-square grid cells were aggregated into 12 square mile polygons, and cell attributes were statistically averaged for each polygon. In the boating survey, boaters were asked to identify their on-the-water destination using a geo-referenced mapping system. Their choice was then linked to the correct polygon with its aggregated site attributes. In this study, we focus on individuals taking day trips.

Statewide there were 26,771 trip-level survey responses during the 12-month sampling period. Of this number, 6690 (25%) reportedly used a boat ramp during their trip. Of those using a boat ramp, 195 (2.9%) used Lee County ramps. Some of these trips used private access (not valid for a public access model), and others failed to identify a valid boat ramp so were removed from the analysis. After adjusting for long distance trips, a total of 153 valid trips were available for the RUM analysis.

Travel costs were computed using the miles traveled from the point of origin to the boat ramp, in addition to the launch fees, which vary by ramp. The miles traveled was derived from the PC-miler software by adding the road miles from the origin of the trip to the location the boat is kept (which are the same in many cases) to the road miles from there to the latitude and longitude geolocation associated with each of the ramp groups. Travel costs were then the sum of the launch fee, bridge tolls, the driving cost assuming towing[B] and the time costs derived as the driving time (m/45 mph) multiplied by the time value (annual income/2080 h per year)[C].

Spatial analysis was conducted with the ArcGIS software, and all numerical analysis was conducted with the LIMDEP software.

5. Results

The estimation results for the model of water site choices, conditional upon a ramp, are presented in Table 1. The table gives the estimated parameters, their standard errors (S.E.), and the levels at which the parameters can be considered to be statistically significant (p-values). The overall model is statistically significant based on a chi-squared test of the joint parameter values. The coefficient of travel cost is significant and of the expected sign.

Table 1. Random utility model estimates for choice of water sites.

Variable (Water Site Characteristic)	Coefficient	S.E.	p-Value
Travel cost	−0.4609	0.0452	<0.0001
Navigation aids in grid	−0.9250	0.4908	0.0595
Artificial reef in grid	−5.1340	2.3967	0.0322
Marine protected or conservation zone in grid	2.1276	0.3721	<0.0001
Manatee zone in grid	−1.2558	0.4550	0.0058
Mean depth	0.3174	0.0672	<0.0001
Nearest ramp distance	−0.4411	0.0904	<0.0001
N = 153			
LLF = −516.65			
McFadden R^2 = 0.209			

The results indicate that the final water destinations chosen by survey respondents are less likely to be in grids with navigation aids (significant at 10% but not at 5%). Similarly, grids with artificial reefs were less likely to be selected as on-the-water destinations. Water sites with marine protected zones or with conservation zones within the grid were significantly more likely to be chosen. Alternatively, water grids with a manatee zone were significantly less likely to be selected as on-the-water destinations. The mean depth of a grid was positively associated with the choice of on-the-water destination, reflecting a preference for deeper water. Finally, the distance from the water site to the nearest ramp (defined as any ramp, not just the ramp they launched from) was negatively associated with the choice of on-the-water destinations. In sum, preferred water destinations had low travel costs, were close to a ramp and near a conservation zone, yet were in deeper water away from navigation aids, artificial reefs, and manatee zones.

The estimation results for the model of ramp site choices are presented in Table 2. The table gives the estimated parameters, their S.E., and the levels at which the parameters can be considered to be statistically significant (p-values). The overall model is significantly based on a chi-squared test of the joint parameter values. The travel cost for getting to the ramp is significant and of the expected (negative) sign, implying that all else equal boaters prefer ramps that are nearer to their points of origin over ramps that are farther away.

Table 2. Random utility model estimates for choice of ramp groups.

Variable (Water Site Characteristic)	Coefficient	S.E.	*p*-Value
Travel cost	−0.0299	0.003	<0.0001
Inclusive value of water sites	0.4586	0.126	0.0003
Number of sites within group	0.8701	0.138	<0.0001
Average parking size (1000's)	0.0328	0.008	0.0001
Parking condition index	0.8340	0.328	0.0111
Ramp development index	4.4716	0.618	<0.0001
Marina	−1.4790	0.237	<0.0001
N = 153			
LLF = −391.25			
McFadden R^2 = 0.281			

The inclusive value parameter for on-the-water sites is significant, and the parameter lies between 0 and 1, which is consistent with the theory for nested logits [2]. The parameter is also significantly different than one, which indicates the superiority of the nesting structure relative to a simple un-nested conditional logit model. The number of ramps within a group was positive and significantly different than zero. The theory of aggregation of sites with random utility models suggests that the number of elements in a group should have a parameter of one [37], and our result is consistent with the aggregation theory since the parameter on the number of ramps in a group is not significantly different from one.

The average parking size is significant and positive, as is the index of parking condition. Ramps with higher levels of development (measured by average facility counts) were significantly preferred to those with lower levels of facilities. However, marinas were less preferred by those trailering their boats to a ramp.

Table 3 presents information for the specific ramp groups. Columns five and six show the survey data on ramp choices (giving both the ramp shares and the frequencies). The seventh column presents the predicted probability of selecting a particular ramp. We can see that the model fit roughly corresponds to the distribution of the sample shares. In particular, the model predicts the highest visitation probability for the ramp with the most visits, and similarly predicts relatively high visitation for sample ramps with high visitation. Similarly, most of the ramps that received low or no visits are predicted to have low probabilities of use. Longitude and latitude coordinates for each ramp group can be found in Appendix A, Table A1.

The second column shows the access value per choice occasion for each of the ramps using the consumer surplus as a welfare measure (Equation (11)), where choice occasion means "taking a trip to a Lee County ramp." This amount represents the lost economic value to boaters if they were to lose access to the site for one choice occasion, yet retain access to the other boat ramps in Lee County. The value is in the range of others reported in the literature but is higher than values reported for access to Hawaii ramps [38]. It is important to note that the values reported in Table 3 are values that accrue to all ramp boating trips made to Lee County (i.e., the scope of choices in the model), and are not the values for a specific visitor that has visited a ramp for which access is lost, which are commonly reported in the literature. In the RUM, we can approximate such site-specific values for lost trips to a specific site by dividing the Lee County values per choice occasion for specific ramps by the probability of making a Lee County trip to that ramp [28,35]. If we make these adjustments for the trips to a particular ramp, we obtain values in the range of $30–40 per trip to a specific ramp, which are within the range of user day values found in the recreation literature.

Table 3. Estimated site values and observed and predicted trips to the ramp groups. Sites with closures due to HABs are indicated. Consumer surplus estimates are reported in 2009 dollars.

Ramp Name	CS Per Choice Occasion	Closure Due to HAB	Lower Bound CS Loss	Survey Data on Ramps		Predicted Probability a Lee County Trip is to a Particular Ramp
				Visitation Shares	Frequency	
BMX Strausser	$1.09	Yes	$1.09	0.00%	0	0.032
Alva Boat Ramp	$0.20	Yes	$0.20	0.00%	0	0.006
Burnt Store Boat Ramp	$1.99	No	$0.00	5.90%	9	0.059
Cape Coral Yacht Basin	$1.64	Yes	$1.64	5.90%	9	0.048
Lovers Key/Carl E. J	$2.71	No	$0.00	9.20%	14	0.07
City of Fort Myers Yacht Basin	$1.11	Yes	$1.11	6.50%	10	0.033
Ft Myers Shores Davis Ramp	$0.34	Yes	$0.34	0.70%	1	0.01
Franklin Locks North	$0.28	Yes	$0.28	0.00%	0	0.008
Franklin Locks South	$0.40	Yes	$0.40	0.70%	1	0.012
Bokeelia Boat Ramp	$0.62	No	$0.00	0.70%	1	0.019
Horton Park	$5.27	Yes	$5.27	9.20%	14	0.144
Imperial River Boat Ramp	$0.42	No	$0.00	3.30%	5	0.012
Koreshan State Historic Site	$0.36	No	$0.00	0.70%	1	0.011
Punta Rassa Boat Ramp	$1.27	Yes	$1.27	9.80%	15	0.037
Sanibel Island	$0.73	Yes	$0.73	2.60%	4	0.022
Bonita Beach Resort Motel	$0.09	No	$0.00	0.00%	0	0.003
Cape Harbour Marina	$0.77	Yes	$0.77	1.30%	2	0.023
Ramp on Ohio Avenue	$0.24	Yes	$0.24	0.00%	0	0.007
Castaways Marina	$1.07	No	$0.00	0.00%	0	0.029
Tween Waters Marina	$1.04	No	$0.00	2.60%	4	0.026
Mullock Creek Marina	$0.28	No	$0.00	5.20%	8	0.008
Fish Trap Marina	$0.20	No	$0.00	0.00%	0	0.006
Riverside Park	$0.11	No	$0.00	0.00%	0	0.003
Pine Island Marina	$0.51	No	$0.00	0.00%	0	0.015
Leeward Yacht Club #2	$0.21	Yes	$0.21	0.00%	0	0.006
Russell Park Ramp	$0.09	Yes	$0.09	0.00%	0	0.003
Burnt Store Marina	$0.17	No	$0.00	1.30%	2	0.005
Pineland Marina	$0.51	No	$0.00	2.00%	3	0.015
Terra Verde Country Club	$0.16	No	$0.00	0.00%	0	0.005
No Name Ramp	$0.21	Yes	$0.21	0.00%	0	0.007
Jug Creek Cottages	$1.49	No	$0.00	7.20%	11	0.044
Monroe Canal Marina	$0.78	Yes	$0.78	5.90%	9	0.023
Viking Marina	$9.15	No	$0.00	19.60%	30	0.236
Hickory Bait and Tackle	$0.36	No	$0.00	0.00%	0	0.011
Inlet Motel	$0.74	No	$0.00	0.00%	0	0.022
TOTAL	$36.61		$14.63			

One caveat for the models we present of Lee County relates to the on-the-water site choice model. Because many of the water site variables are spatially correlated (due to the fact many on-the-water sites are in the same 12 square mile polygon), the model is not well suited to evaluating the effect of changes in individual on-the-water site characteristics. However, the model does perform well in terms of predicting water site choice, and hence, the model does a good job of predicting the utility index (inclusive value) of the available on-the-water sites from any ramp. Thus, the combined models are well suited to the valuation of ramps, but less-well suited to the valuation of changes in specific water site characteristics.

The model we present is based on boaters that have launched from ramps in Lee County. Thus, the scope of the model or what might be referred to as the "market area" covered by the model is boaters utilizing public ramps in Lee County. Lee County is a large area with many possible public ramps available to boaters. It is natural to think that ramps within Lee County are a part of the relevant market area for the segment of boaters that have used a Lee County ramp. These ramps are also natural substitute sites for Lee County boaters. Our model includes these possibilities. However, it may be that the geographic market area includes some ramps and boaters using other ramps outside of Lee County. For example, when the characteristics of a Lee County ramp are improved, it may attract some boaters that were not previously using a Lee County ramp. These boating behaviors occurring outside of Lee County would not be captured by these Lee County RUMs. In this case, our model may underestimate the benefits of a Lee County ramp improvement because it cannot capture the benefits to potential new users of Lee County ramps. That said, when improvement occurs, we know that the main beneficiaries are those already using Lee County ramps and these benefits are captured by our models. Conversely, our model will overstate losses due to ramp closures because boaters cannot adjust their trips by going outside of Lee County or by reducing their total trips. This effect will be small for closures affecting sites with low visitation or sets of sites with relatively low visitation [35].

Simulated Losses Due to HAB-Related Ramp Closures

Given the noxious nature of cyanobacterial HABs, losses resulting from these events can be simulated as closures of sites that are impacted by blooms. To create a realistic scenario of site closures during a cyanobacterial HAB, we identify the boat ramps located in the low salinity Caloosahatchee River and differentiate them from those located in high salinity areas beyond the mouth of the river (Figure 1). Given cyanobacterial biology, HABs caused by these organisms can be expected to dissipate as the blooms reach high salinity areas. Therefore, HABs can be expected to cause actual or perceived loss of access from boat ramps located in the river, but access will remain relatively untouched in ramps located in the Gulf of Mexico or in high salinity areas beyond the river mouth.

Therefore, losses accruing to recreational boaters as a result of cyanobacterial HABs in Lee County can be estimated as the difference between the CS with the full choice set and the CS with the restricted choice set where the boat ramps in the river are assumed to be closed (Equation (11)). To approximate this, we can sum up the losses associated with individual site closures. As shown in Table 3, lost value due to closure can be measured by the sum of the lost CS per choice occasion over the full choice set, which was estimated to be $14.63 (2009 dollars, or $17.26 2018 dollars). This approximation undervalues the loss because it does not account for the fact that multiple sites were closed at once nor does it include any losses in fixed costs during short-term closures [8], yet the approximation could overvalue the losses since the model does not allow for substitution out of boating or out of Lee County. Nevertheless, the approach does illustrate how to easily derive a reasonable estimate of losses if one does not have access to the underlying data needed to compute Equation (11). In other words, the average individual who would have taken a boating trip to Lee County on a day when cyanobacterial HABs were occurring in 2018 is estimated to have lost $17.26 as a direct result of the algae blooms.

An aggregate loss for the months when HABs occur can be estimated by multiplying the $17.26 per choice occasion loss in consumer surplus as a result of the cyanobacterial HABs by the number of Lee County trips that would normally have occurred in those months if there were no HABs. Table 4 provides aggregate loss estimates for each month of the year, assuming that the algae bloom lasts for the entire duration of the month. These monthly figures take into account seasonal recreation patterns and expected annual growth in trips (Figure 4, [9]).

Table 4. Consumer surplus lost in closure scenarios for 12 months (2018 dollars).

Month of Closure	Estimated Loss
January	$729,752.64
February	$624,098.22
March	$653,583.17
April	$928,776.09
May	$948,432.72
June	$938,604.41
July	$992,660.16
August	$1,041,801.75
September	$1,592,187.58
October	$1,253,110.59
November	$914,033.61
December	$835,407.06

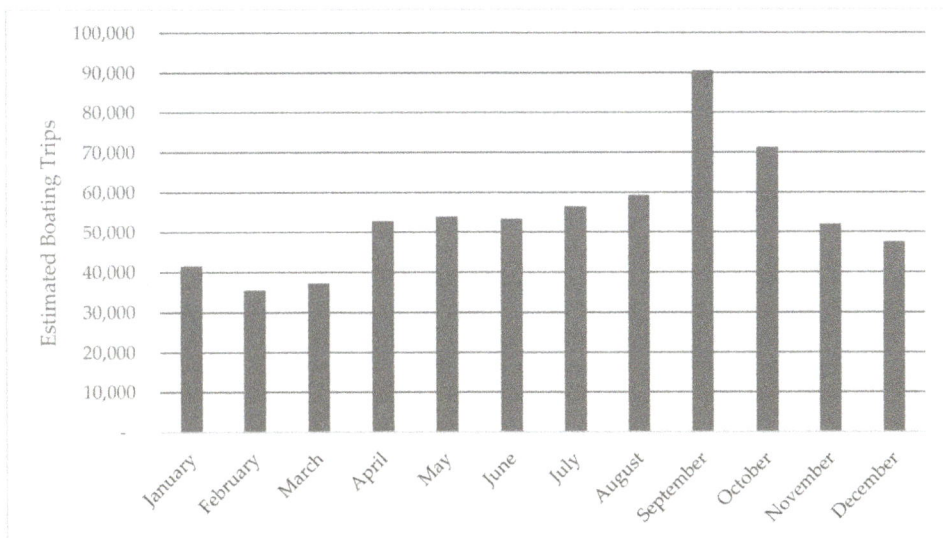

Figure 4. Estimated monthly number of boating trips to Lee County. Adapted from FWC [9].

This framework can be used to estimate losses from historic or hypothetical algae blooms in Lee County. For example, the 2018 blooms were first documented at the end of June, and relatively high concentrations of the cyanotoxin microcystin were documented through July, August, and September, with relatively lower concentrations documented all the way to the end of the year [27]. Thus, losses for the 2018 cyanobacterial HAB in Lee County can be estimated at $3,550,537 (2018 dollars), or the sum of the monthly estimates for July, August, and September. Losses for hypothetical (or observed) HAB scenarios of different durations can be estimated using Table 4 along with the hypothetical (or observed) duration of the HABs.

6. Discussion

While the model provides credible estimates of the losses to recreational boaters as a result of the cyanoHABs, it is limited in several accounts. Estimates produced by this model are biased in both directions. On one hand, the model underestimates since (1) the site-specific CS estimates in Table 3 do not jointly evaluate the closure, and, since (2) they do not account for any losses in fixed costs due to short-term closures [8]. On the other hand, the model overestimates since (1) discontinuing participation in Lee County boating is not a substitution option in the model, and since (2) there may be days where HABs cause a decline in site quality but do not fully close a ramp. The assumption that HABs result in full closure is similar to the assumption made by [39] to illustrate how to use benefit transfer to estimate damages of HABs on Lake Erie.

The per choice occasion welfare loss of $17.26 reported in this study is in the range of others previously estimated and reported in the literature. For instance, [32] has shown that when reviewing the effects of the BP oil spill in the Gulf of Mexico, losses have been reported as low as $2.23 for individuals fishing from private boats, but as high as $34.27 for individuals fishing from chartered boats. Similarly, [33] reports welfare estimated losses to shoreline recreationists of between $37.23 and $40.41 per lost trip as a result of the BP oil spill.

It is important to note that while these results provide an estimate of some losses in ecosystem services arising from HABs, they do not provide a complete picture of all losses in ecosystem services resulting from these events. Our estimates only include losses to recreational boaters, and do not account for losses in other cultural, provisioning, regulating or supporting ecosystem services that are impacted by cyanobacterial HABs. Thus, losses from HABs in Lee County can be expected to be larger than the estimates provided here.

Future research efforts aimed at better understanding the impact of HABs on recreational services of coastal and marine ecosystems could focus on estimating the demand response to HABs that are near a ramp but do not close a ramp. These estimates can be obtained using stated preference surveys [12] or by developing estimates based on empirical observations of changes in behavior resulting from changes in the quality of ecosystem services [32,33], rather than on simulations using previously collected data (such as the estimates developed in this study).

The approach developed in this study could be coupled with biophysical models of the movement of HABs or other hazards that negatively impact the provision of cultural ecosystem services to develop quasi-real-time damage estimates as a result of these hazards. For instance, a biophysical model of the movement of HABs along the Caloosahatchee River could be used in tandem with the site- or ramp-based approach developed here to simulate closures in individual boat ramps as the cyanobacterial blooms flow from Lake Okeechobee to the Gulf of Mexico. Similar site-based closure simulations could be applied to other settings where movement of hazards is relatively predictable.

7. Conclusions

A RUM model was used to compute the value of changing site characteristics as well as to estimate the value of access for available recreational boating sites in Lee County. As expected, the more popular ramps have higher per choice occasion values, but all sites have fairly similar values per-trip to the site. The model was also applied to assess the loss of cultural ecosystem services in terms of foregone recreational boating opportunities as a result of harmful algae blooms. Losses to recreational boating resulting from the 2018 blooms in Lee County are estimated at $3.5 million (2018 dollars). This approach can be adapted to other contexts where ecological disturbances are causing closures of recreational areas.

A. The Millennium Ecosystem Assessment (MEA 2005) defines cultural ecosystem services as nonmaterial benefits that are enjoyed through recreation and aesthetic experiences, spiritual or artistic appreciation.

B. Travel costs were assumed to be $0.50 per mile in 2009 dollars (equivalent to $0.59 2018 dollars).

C. Travel times for two sites (Sanibel and Lovers Key) were adjusted downward to 20 mph for a portion of their travel distance to account for slower speeds on causeways and highly congested areas.

Supplementary Materials: The following are available online at http://www.mdpi.com/2073-4441/11/6/1250/s1, Data: KML file with boat ramp locations that indicates whether they are open or closed in our HAB scenario.

Author Contributions: Conceptualization, S.A., F.L., D.S., M.T.; methodology, S.A., F.L., D.S., M.T.; software, F.L.; validation, F.L., M.T.; formal analysis, S.A., F.L.; investigation, F.L., M.T.; resources, S.A., F.L., D.S., M.T.; data curation, F.L, M.T.; writing—original draft preparation, S.A., D.S.; writing—review and editing, S.A., F.L., D.S., M.T.; visualization, S.A.; supervision, S.A., F.L., D.S., M.T.; project administration, M.T.; funding acquisition, M.T.

Funding: The initial phase of this research was funded by the Florida Fish and Wildlife Conservation Commission, grant number FWC 04/05-23.

Acknowledgments: The support of the Florida Fish and Wildlife Conservation and Lee County are gratefully acknowledged. We thank Dave B. Harding and Ed Mahoney for their input in the initial phase of research.

Conflicts of Interest: The authors declare no conflict of interest. The funders had no role in the design of the study; in the collection, analyses, or interpretation of data; in the writing of the manuscript, or in the decision to publish the results.

Appendix A

Table A1. Latitude and longitude coordinates of boat ramp groups in the study.

Ramp Name	Lat.	Long.
BMX Strausser	26.62438	−81.98989
Alva Boat Ramp	26.71398	−81.60597
Burnt Store Boat Ramp	26.64714	−82.04193
Cape Coral Yacht Basin	26.54266	−81.95226
Lovers Key/Carl E. J	26.39354	−81.86658
City of Fort Myers Yacht Basin	26.64579	−81.87249
Ft Myers Shores Davis Ramp	26.71155	−81.75297
Franklin Locks North	26.72375	−81.69232
Franklin Locks South	26.72098	−81.69445
Bokeelia Boat Ramp	26.69431	−82.1459
Horton Park	26.60736	−81.91359
Imperial River Boat Ramp	26.33863	−81.80495
Koreshan State Historic Site	26.43663	−81.81962
Punta Rassa Boat Ramp	26.48458	−82.01049
Sanibel Island	26.45354	−82.03592
Bonita Beach Resort Motel	26.35391	−81.85499
Cape Harbour Marina	26.54383	−82.00788
Ramp on Ohio Avenue	26.45115	−81.94748
Castaways Marina	26.48152	−82.18026
Tween Waters Marina	26.5109	−82.18947
Mullock Creek Marina	26.47347	−81.8516
Fish Trap Marina	26.3306	−81.83171
Riverside Park	26.34232	−81.77971
Pine Island Marina	26.59273	−82.12625
Leeward Yacht Club #2	26.68775	−81.79311
Russell Park Ramp	26.68295	−81.81462
Burnt Store Marina	26.76042	−82.05518
Pineland Marina	26.66218	−82.15527
Terra Verde Country Club	26.49037	−81.85468
No Name Ramp	26.67372	−81.90007
Jug Creek Cottages	26.70388	−82.15735
Monroe Canal Marina	26.50525	−82.08285
Viking Marina	26.6354	−82.06264
Hickory Bait and Tackle	26.39939	−81.84127
Inlet Motel	26.7311	−82.21302

References

1. US Coast Guard. *Recreational Boating Statistics*; US Department of Homeland Security: Washington, DC, USA, 2017.
2. Morey, E. TWO RUMs unCLOAKED: Nested-Logit Models of Site Choice and Nested-Logit Models of Participation and Site Choice. In *Valuing the Environment Using Recreation Demand Models*; Kling, C.L., Herriges, H., Eds.; Edward Elgar Publishing Ltd.: Northhampton, MA, USA, 1999; Chapter 4.
3. Milon, J. A Nested Demand Shares Model of Artificial Marine Habitat Choice by Sport Anglers. *Mar. Resour. Econ.* **1988**, *5*, 191–213. [CrossRef]
4. Bockstael, N.; McConnell, K.; Strand, I. A Random Utility Model of Sport Fishing: Some Preliminary Results for Florida. *Mar. Resour. Econ.* **1989**, *6*, 245–260. [CrossRef]
5. Morey, E.; Rowe, R.; Watson, M. A Repeated Nested Logit Model of Atlantic Salmon Fishing. *Am. J. Agric. Econ.* **1993**, *75*, 578–592. [CrossRef]
6. Greene, G.; Moss, C.; Spreen, T. Demand for Recreational Fishing in Tampa Bay, Florida: A Random Utility Approach. *Mar. Resour. Econ.* **1997**, *12*, 293–305. [CrossRef]
7. Thomas, M.; Stratis, N. Compensating Variation for Recreational Policy: A Random Utility Approach to Boating in Florida. *Mar. Resour. Econ.* **2002**, *17*, 23–33. [CrossRef]
8. English, E.; Herriges, J.; Lupi, F.; McConnell, K.; von Haefen, R. Fixed costs and recreation value. *Am. J. Agric. Econ.* **2019**. [CrossRef]

9. Florida Fish and Wildlife Conservation (FWC). *The Florida Boating Access Facilities Inventory and Economic Study Including A Pilot Study for Lee County*; RFP NO. FWC 04/05-23; Florida Fish and Wildlife Conservation Commission: Tallahassee, FL, USA, 2009.

10. Carmichael, W.W. Health Effects of Toxin-Producing Cyanobacteria: "The CyanoHABs". *Hum. Ecol. Risk Assess. Int. J.* **2001**, *7*, 1393–1407. [CrossRef]

11. Zhang, F.; Lee, J.; Liang, S.; Shum, C.K. Cyanobacteria blooms and non-alcoholic liver disease, evidence from a county level ecological study in the United States. *Environ. Health* **2015**, *14*, 1–11. [CrossRef]

12. Zhang, W.; Sohngen, B.; Do, U.S. Anglers Care about Harmful Algal Blooms? A Discrete Choice Experiment of Lake Erie Recreational Anglers. *Am. J. Agric. Econ.* **2018**, *100*, 868–888. [CrossRef]

13. Willis, C.; Papathanasopoulou, E.; Russel, D.; Artioli, Y. Harmful algal blooms: The impacts on cultural ecosystem services and human well-being in a case study setting, Cornwall, UK. *Mar. Policy* **2018**, *97*, 232–238. [CrossRef]

14. Garcia-Ayllon, S. The Integrated Territorial Investment (ITI) of the Mar Menor as a model for the future in the comprehensive management of enclosed coastal seas. *Ocean Coast. Manag.* **2018**, *166*, 82–97. [CrossRef]

15. Dyson, K.; Huppert, D.D. Regional economic impacts of razor clam beach closures due to harmful algal blooms (HABs) on the Pacific coast of Washington. *Harmful Algae* **2010**, *9*, 264–271. [CrossRef]

16. Backer, L.C. Impacts of Florida red tides on coastal communities. *Harmful Algae* **2009**, *8*, 618–622. [CrossRef]

17. Hoagland, P.; Jin, D.; Beet, A.; Kirkpatrick, B.; Reich, A.; Ullmann, S.; Fleming, L.E.; Kirkpatrick, G. The human health effects of Florida Red Tide (FRT) blooms: An expanded analysis. *Environ. Int.* **2014**, *68*, 144–153. [CrossRef] [PubMed]

18. Morgan, K.L.; Larkin, S.L.; Adams, C.M. Red tides and participation in marine-based activities: Estimating the response of Southwest Florida residents. *Harmful Algae* **2010**, *9*, 333–341. [CrossRef]

19. Visit Florida 2017–2018 Year in Review. Available online: https://www.visitflorida.org/media/57255/yearinreview2018.pdf (accessed on 1 May 2019).

20. Havens, K. Watching and Waiting: Uncertainty About When Algae Blooms Will End. *Florida Sea Grant*. Available online: https://www.flseagrant.org/news/2018/10/watching-and-waiting-uncertainty-about-when-algae-blooms-will-end/ (accessed on 1 May 2019).

21. Dodds, W.K.; Bouska, W.W.; Eitzmann, J.L.; Pilger, T.J.; Pitts, K.L.; Riley, A.J.; Schloesser, J.T.; Thornbrugh, D.J. Eutrophication of US freshwaters: Analysis of potential economic damages. *Environ. Sci. Technol.* **2009**, *43*, 12–19. [CrossRef]

22. Adams, C.M.; Larkin, S.L.; Hoagland, P.; Sancewich, B. Assessing the economic consequences of harmful algal blooms: A summary of existing literature, research methods, data, and information gaps. In *Harmful Algal Blooms: A Compendium Desk Reference*; Shumway, S.E., Burkholder, J.M., Morton, S.L., Eds.; John Wiley and Sons: Hoboken, NJ, USA, 2018.

23. Heisler, J.; Glibert, P.M.; Burkholder, J.M.; Anderson, D.M.; Cochlan, W.; Dennison, W.C.; Dortch, Q.; Gobler, C.J.; Heil, C.A.; Humphries, E.; et al. Eutrophication and harmful algal blooms: A scientific consensus. *Harmful Algae* **2008**, *8*, 3–13. [CrossRef]

24. Grunwald, M. Everglades. *Smithson. Mag.* **2006**, *36*, 47–57.

25. Rosen, B.H.; Loftin, K.A.; Graham, J.L.; Stahlhut, K.N.; Riley, J.M.; Johnston, B.D.; Senegal, S. *Understanding the Effect of Salinity Tolerance on Cyanobacteria Associated with A Harmful Algal Bloom in Lake Okeechobee, Florida*; U.S. Geological Survey Scientific Investigations Report; US Department of the Interior: Reston, VA, USA, 2018.

26. Tonk, L.; Bosch, K.; Visser, P.M.; Huisman, J. Salt tolerance of the harmful cyanobacterium Microcystis aeruginosa. *Aquat. Microb. Ecol.* **2007**, *46*, 117–123. [CrossRef]

27. Algal Bloom Sampling Results. Available online: https://floridadep.gov/dear/algal-bloom/content/algal-bloom-sampling-results (accessed on 1 May 2019).

28. Freeman, A.M.; Herriges, J.A.; Kling, C.L. *The Measurement of Environmental and Resource Values: Theory and Methods*, 3rd ed.; RFF Press: Washington, DC, USA, 2014.

29. Parsons, G.R. Travel cost models. In *A Primer on Nonmarket Valuation*, 2nd ed.; Champ, P.A., Boyle, K.J., Brown, T.C., Eds.; Springer: Dordrecht, The Netherlands, 2017.

30. Tomasi, T.; Thomas, M. *Natural Resource Damage Assessment for the Tampa Bay Oil Spill: Recreational Use Losses for Florida Residents*; Draft Report; Florida Department of Environmental Protection: Tallahassee, FL, USA, 1998.

31. Lupi, F.; Hoehn, J.; Christie, G. Using an Economic Model of Recreational Fishing to Evaluate Benefits of Sea Lamprey (Petromyzon marinus) Control. *J. Great Lakes Res.* **2003**, *29*, 742–754. [CrossRef]

32. Alvarez, S.; Larkin, S.L.; Whitehead, J.C.; Haab, T.C. A revealed preference approach to valuing non-market recreational fishing losses from the Deepwater Horizon oil spill. *J. Environ. Manag.* **2014**, *145*, 199–209. [CrossRef] [PubMed]

33. English, E.; von Haefen, R.H.; Herriges, J.; Leggett, C.; Lupi, F.; McConnell, K.; Welsh, M.; Domanski, A.; Meade, N. Estimating the Value of Lost Recreation Days from the Deepwater Horizon Oil Spill. *J. Environ. Econ. Manag.* **2018**, *91*, 26–45. [CrossRef]

34. Florida Fish and Wildlife Conservation (FWC). *Schedule of Estimated Regulatory Cost: Brevard County Manatee Protection Rule*; Florida Fish and Wildlife Conservation Commission: Tallahassee, FL, USA, 2002.

35. Haab, T.; McConnell, K. *Valuing Environmental and Natural Resources: The Econometrics of Nonmarket Valuation*; Edward Elgar: Northhampton, MA, USA, 2002.

36. Small, K.; Rosen, H. Applied Welfare Economics with Discrete Choice Models. *Econometrica* **1982**, *49*, 105–130. [CrossRef]

37. Lupi, F.; Feather, P. Using Partial Site Aggregation to Reduce Bias in Random Utility Travel Cost Models. *Water Resour. Res.* **1998**, *34*, 3595–3603. [CrossRef]

38. Haab, H.; Hamilton, M.; McConnell, K. Small Boat Fishing in Hawaii: A Random Utility Model of Ramp and Ocean Destination. *Mar. Resour. Econ.* **2008**, *23*, 137–151. [CrossRef]

39. Palm-Forster, L.; Lupi, F.; Chen, M. Valuing Lake Erie beaches using value and function transfers. *Agric. Resour. Econ. Rev.* **2016**, *45*, 270–292. [CrossRef]

![water logo] *water*

MDPI

Article

Coastal and Marine Quality and Tourists' Stated Intention to Return to Barbados

Peter Schuhmann [1,*], Ryan Skeete [2], Richard Waite [3], Prosper Bangwayo-Skeete [4], James Casey [5], Hazel A. Oxenford [6] and David A. Gill [7]

[1] Department of Economics and Finance, University of North Carolina Wilmington, Wilmington, NC 28403, USA

[2] Caribbean Tourism Organization, Warrens, St. Michael BB22026, Barbados; skeeteryan@gmail.com

[3] World Resources Institute, Washington, DC 20002, USA; RWaite@wri.org

[4] Department of Economics, The University of the West Indies, Cave Hill BB11000, Barbados; prosper.bangwayo-skeete@cavehill.uwi.edu

[5] Williams School of Commerce, Economics and Politics, Washington and Lee University, Lexington, VA 24450, USA; caseyj@wlu.edu

[6] Centre for Resource Management and Environmental Studies, University of the West Indies, Cave Hill BB11000, Barbados; oxenford.hazel@gmail.com

[7] Nicholas School of the Environment, Duke University Marine Lab, Beaufort, NC 28516, USA; david.gill@duke.edu

* Correspondence: schuhmannp@uncw.edu; Tel.: 1-910-962-3417

Received: 2 May 2019; Accepted: 13 June 2019; Published: 17 June 2019

Abstract: Seawater quality is critical for island and coastal communities dependent on coastal tourism. Improper management of coastal development and inland watersheds can decrease seawater quality and adversely impact marine life, human health, and economic growth. Agricultural runoff and improper sewage management compromise nearshore water quality in many coastal regions and can impact visitation decisions of tourists who are drawn to these destinations. The purpose of this paper is to understand how tourists' decisions to revisit Barbados might be affected by changes in coastal and marine quality. We use data collected from tourists to examine how tourists' stated willingness to return is affected by scenarios involving changes in seawater quality, beach width and coral reef health. Results reveal that return decisions are sensitive to changes in all aspects of coastal and marine quality. A reduction in seawater quality discourages tourists' intention to return more than other environmental factors. These results are of paramount interest to destination managers, marketers and policymakers who rely on repeat visitation data to develop marketing strategies and infer future direction. This research highlights the importance of prioritizing seawater quality management to protect the coastal tourism product, especially in small island developing states (SIDS) with a high reliance on tourism income.

Keywords: seawater quality; contingent behavior; tourism; Barbados

---

1. Introduction

Clean beaches, clear turquoise waters and healthy coral reefs provide the principal settings for tourism activity in the Caribbean [1,2]. Coastal development is needed to support this tourism, but when improperly managed can have negative impacts on the natural assets that attract visitors. For example, losses in coral reef quality can have an array of impacts due to their prominent role in creating and protecting beaches, protecting coastal infrastructure, and providing aesthetic value for divers and snorkelers.

Management of coastal runoff and sewage is especially important for areas that depend on coastal tourism, as seawater quality is directly connected to human and ecosystem health. Approximately

30 percent of the Caribbean's coral reefs are at high risk from threats like runoff, sedimentation and the discharge of untreated domestic and hotel wastewater [3,4]. When nutrient-rich effluents found in agricultural runoff and sewage are deposited into natural water courses [5], microalgae (phytoplankton) [6] and macroalgae (seaweed) grow rapidly and out-compete coral for space on the reef [7]. Suspended sediment also blocks light [6], and when it settles it smothers the reef, destroying living spaces, damaging corals and other benthic fauna and preventing settlement of new coral recruits [8]. Contaminated seawater can also lead to a variety of human health hazards through ingestion or contact with pathogenic microorganisms, including gastroenteric illnesses such as diarrhea, or infections of the upper respiratory tract, ears, eyes, nasal cavity and skin [9]. Real or perceived risks involving the quality of recreational seawater therefore have important economic implications in areas that depend on tourism as a primary source of income [10].

A significant and growing literature attempts to understand the tradeoffs between coastal and marine quality and the economic returns from tourism. Much of this literature uses revealed and stated preference valuation methods to estimate willingness to pay (WTP) for "non-market" aspects of coastal and marine quality such as beach width, coral reef health and biodiversity. WTP estimates can be informative for policy and management decisions. For example, estimates of WTP can be used to understand the economic benefits of improving or maintaining resource quality to inform cost-benefit analyses. WTP estimates can also be used to identify potential sources of funding for conservation and management, to design incentives that promote sustainable behavior, and to highlight the opportunity costs of resource degradation.

Because the assumptions necessary to employ revealed preference methods do not apply to many non-market environmental goods [11], and actual (behavioral) data can be costly to obtain, especially for goods that are purchased infrequently [12], stated preference methods such as the contingent valuation method (CVM) and choice experiments (CE) are often employed in WTP investigations. Stated preference methods provide the advantage of allowing for the examination of scenarios involving hypothetical changes that are outside the range of historical conditions [13] and can be used to estimate non-use values.

Recent examples include Halkos and Matsiori [14], who use CVM to estimate residents' WTP for a state-managed program of coastal zone improvements in Volos, Greece. Their results suggest that a substantial proportion of respondents are willing to pay for improving coastal zone quality, and that improving the quality of bathing water is the most important reason for WTP. García-Ayllón [15] present a novel approach to CVM, combining WTP estimates with participatory GIS mapping and process optimization to formulate environmental management solutions on the Spanish Mediterranean coast. Recent applications of CVM in the Caribbean include Casey and Schuhmann [16] and Schuhmann et al. [17], who measure tourists' willingness to pay conservation fees in Belize and Barbados respectively, and Trujillo et al. [18], who use the contingent valuation method (CVM) to estimate recreational divers' willingness to pay for the conservation of the coral reefs in the Columbian Caribbean.

Applications of the CE method include Pakalniete et al. [19], who estimate Latvian citizens' WTP for improving seawater quality for recreation, avoiding reductions in marine biodiversity and limiting new occurrences of invasive alien species in coastal and marine waters of the Baltic Sea. Of these attributes, they find that WTP is highest for better seawater quality. Christie et al. [20] estimate the benefits from six attributes associated with marine protected areas (MPAs) in St. Vincent and the Grenadines (SVG): Fishing quality, coastal protection, water quality (as related to human health), species diversity, beach recreation and diving/snorkeling recreation. The authors find that both locals and tourists on the South coast of St. Vincent were willing to pay the most for avoiding declines in human health related to water quality, followed by improvements to human health. Numerous other WTP studies estimate the monetary value that Caribbean beachgoers and underwater recreationists attribute to clean and wide beaches, healthy reefs, diverse fish populations and encounters with species such as sea turtles [21–23], with similar results found outside the region [24–26].

While monetary estimates of WTP for changes in resource quality are valuable for many reasons, policy makers in tourism-dependent areas can also benefit from an appreciation for how people's behavior will change in response to changes in resource quality. For example, an understanding of how tourists' willingness to return to a destination is related to the quality of the natural environment can help policy makers plan for changes in demand, anticipate changes in visitor profiles or plan targeted interventions to influence the behavior of demographic groups [27]. In tourism-dependent destinations where arrivals and return visitation are important measures of success, understanding how willingness to return depends on environmental quality might be a more tangible and relevant measure of socioeconomic outcomes than WTP. Furthermore, this "contingent behavior" approach may minimize some of the difficulties associated with traditional WTP estimation, such as respondents' unfamiliarity with formulating monetary estimates of WTP, and moral objection to monetizing nature. While methodologically similar to CVM, applications of the contingent behavior approach are relatively rare in the literature, especially as related to Caribbean tourism.

The purpose of this study is to complement and add to the existing valuation literature by assessing the potential effect of changes in seawater quality, beach width and coral reef health on visitors' stated intention to return to Barbados. This inquiry is novel, as we estimate Caribbean tourists' willingness to return to a destination under conditions of environmental change, rather than willingness to pay, thereby complementing existing WTP studies and adding an important quantity-based measure to our understanding of tourism demand. Our results can help guide policy by providing insight into how continued environmental degradation might affect return visitation and how changes in resource quality might alter the sociodemographic profile of the Barbados tourist population.

2. Materials and Methods

2.1. Study Site

Located in the southeastern Caribbean Sea, tourism is the leading foreign exchange earner in Barbados. In 2018, the tourism sector accounted for more than 40 percent of Barbados GDP and national employment [28] and 60.8 percent of total exports. In 2017, Barbados attracted 661,160 overnight tourist arrivals, more than twice its population [29], and a record 681,211 cruise ship passengers [30].

Barbados is among the top-ranking countries for dependence on reefs, yet also among the top countries for highest exposure of the reef to threats [4]. Live coral cover and reef community health have declined considerably since the 1970s [31–33]. These declines have been linked to increasing seawater eutrophication and wastewater pollution from hotels and coastal properties [8,34–39]. Barbados policymakers were prompted to engage in numerous measures, including constructing a 44-km central primary sewage system to capture wastewater flows on the south coast (where hotels are spatially concentrated), implementing water quality standards, forming an oversight committee for water pollution control, and offering tax incentives for upgrades to hotel wastewater disposal systems [6].

Notwithstanding these attempts to translate policy into action, weaknesses remain. The level of sewage treatment in Barbados remains inadequate to protect nearshore ecosystems [6,40], thereby threatening the vital economic contribution of the tourism sector. Indeed, in 2017–2018, a failure at the island's south coast sewage treatment plant resulted in the release of untreated sewage in an area of high tourism activity. As a result, Barbados' main source markets—USA, U.K., Canada and Germany—issued health alerts and travel advisories against visiting Barbados' south coast. Due to health concerns, the government closed Worthing Beach for extended periods from December 2016 to December 2018. Hotels and guest houses in the area have lost business, and several restaurants and shops closed, as many tourists relocated or cancelled reservations.

Policymakers are keenly aware of these problems, but are also concerned about the costs associated with increased waste-water treatment, including disruptions to tourism, transport and local commerce [41]. The costs of increasing and maintaining waste-water treatment facilities can be significant, yet are easily calculable for planning purposes. However, because the benefits associated with improved seawater quality and the costs of continued losses have not been quantified, policy makers lack sufficient information to evaluate the economic implications of management interventions or the consequences of inaction. It is this gap in knowledge that this research seeks to address.

2.2. Survey

Similar to the approach taken by Hanley et al. [42] and based on prior work in Barbados by Schuhmann et al. [17,43], an exit questionnaire was developed to collect a variety of trip and individual characteristics from departing visitors to Barbados. In addition to providing an array of demographic information such as age, income, country of origin and education level, respondents provided information about their trip, including trip purpose, lodging type, participation in recreational activities in the coastal zone and Likert scale ratings of several coastal and marine attributes.

Early in the survey, respondents were asked to state the likelihood that they would return to Barbados, choosing from "definitely", "probably", "unsure", "probably not" and definitely not". Later in the survey respondents were (again) asked four separate questions about their willingness to return to Barbados, each pertaining to a scenario describing a change in environmental conditions. Specifically, respondents were asked to state the likelihood that they would return to Barbados, choosing from the same responses noted above, if changes occurred to beach width, coral reef health, the quality of marine life (fish, turtles, etc.), and the cleanliness of the sea water. All environmental changes were described in terms of percentage changes from the current condition. The specific text used in the questionnaire is shown in Table 1 below, where we also include corresponding variable names and the coding scheme which ranged from 1–5. The strongest negative response (definitely will not return) was coded as 1 and the strongest positive response (definitely will return) was coded as 5.

Table 1. Willingness to return survey questions.

Variable Name	Survey Question (10 Percent Reduction in Quality)
Plan_Return	8. Do you plan to return to Barbados in the future?
Plan to return if beach width changes (Plan_Return_BW)	21. If beach widths in Barbados were to decrease by 10%, and all other conditions remained the same, would you return to Barbados in the future?
Plan to return if coral health changes (Plan_Return_CH)	22. If coral reef health in Barbados was to decline by 10%, and all other conditions remained the same, would you return to Barbados in the future?
Plan to return if marine life changes (Plan_Return_ML)	23. If the quality of marine life (fish, turtles, etc.) in Barbados were to decrease by 10%, and all other conditions remained the same, would you return to Barbados in the future?
Plan to return if seawater quality changes (Plan_Return_SW)	24. If the cleanliness of sea water in Barbados were to change so that your risk of a stomach infection increased by 10%, and all other conditions remained the same, would you return to Barbados in the future?
Responses (data code)	Definitely not (1) Probably not (2) Unsure (3) Probably (4) Definitely (5)

Sixteen different versions of the survey were randomly distributed among respondents. Each version included one of seven different values of hypothetical environmental changes. Three of these changes were improvements (5, 10 and 25 percent increases in beach width, coral reef health, marine life and seawater quality) and four described degraded environmental conditions (5, 10, 25 and 50 percent reductions). Due to current trends in Barbados, degradations were deemed more likely than improvements. We therefore weighted the distribution of environmental change parameters toward degradations. Roughly 70 percent of respondents were presented with scenarios involving negative changes. Because we used seven values of the environmental change parameter and 16 versions of the survey, the change parameters were not evenly distributed across the sample. For example, −10 and −25 percent changes were presented to 25 percent and 20 percent of the respondents respectively, while the scenario involving 5 percent improvements was presented to 5.5 percent of the sample.

Seawater quality can be expressed in terms of water clarity or visibility [44,45], meeting regulatory standards of quality [42,46], the chance of human illness through contact [47–49] or the presence of algae [44,47]. Water-based recreation can expose individuals to a variety of health hazards arising from contamination by runoff, sewage and excreta [9,49]. According to the World Health Organization [9], enteric illness is the most frequent adverse health outcome associated with exposure to fecally contaminated recreational water. Pond [10] notes that gastroenteric symptoms such as diarrhea are widespread and common among recreational water users. Expressing changes in the quality of seawater in terms of the risk of a stomach infection relative to the status quo condition was judged to be the most appropriate for policy decisions in Barbados.

2.3. Data

The questionnaire was distributed to visitors in the departures lounge of the Grantley Adams International Airport (GAIA) in the final two weeks of March 2015. The data were collected prior to the negative publicity surrounding the sewage issues and are a subset of the data used in the Schuhmann et al. analysis of tourists' willingness to pay conservation fees [17]. After removal of incomplete questionnaires and questionnaires that included extreme outliers or illogical responses, approximately 3300 completed questionnaires were retained.

We focus on how responses to the willingness to return questions changed from the first version of the question (status quo conditions, prior to mention of environmental change) to the scenarios that described changes in coastal and marine quality. Because visitation decisions by business travelers (roughly 8 percent of the sample) and Barbadian nationals living abroad (4.7 percent) are likely to be based on non-environmental factors, we removed these two classes of individuals from the sample. Approximately 2700 of the remaining respondents answered both the initial willingness to return question and at least one question regarding willingness to return with changed environmental conditions. From this subsample we remove observations where the respondent initially suggested that they would (would not) return to Barbados, and subsequently indicated a higher probability that they would (would not) return under degraded (improved) environmental conditions. We assume that these respondents did not understand at least one of the willingness to return questions. Our final sample for analysis contains approximately 2550 observations.

2.4. Empirical Approach

We use different forms of general linear modeling (GLM) and logit regression to identify factors associated with respondents changing their stated willingness to return to Barbados under status quo conditions (Question 8, Table 1) after being presented with hypothetical changes in environmental quality (Questions 22–24). Changes in the stated intention to return can be measured several ways. First, we can quantify the difference in responses between the two intention to return questions as a single numerical value. For example, a "change in score" measure for changes in intention to revisit following the scenario describing changes in environmental quality can be measured using a new variable, Y_3:

$$Change\_Intention = Y_3 = Y_2 - Y_1 \tag{1}$$

where Y_1 is the response to the initial willingness to return question and Y_2 is the response to the subsequent question under conditions of environmental change. This variable can be constructed for changes in water quality, beach width, coral reef health and the quality of marine life, and is an ordinal variable with nine theoretically possible outcomes ranging from −4 to 4. Negative values of *Change_Intention* correspond to potential losses (respondents who are less likely to return), while positive values represent potential gains in return visitation (respondents who are more likely to return). A larger absolute value indicates a higher magnitude of change from the initial stated response. Table 2, below, shows the possible composition of values for *Change_Intention* outcomes.

Table 2. Composition of values for *Change_Intention* Score.

		Change_Intention_XX = Plan_Return_XX [a] − Plan_Return		
	−4	Definitely No − Definitely Yes		
Losses	−3	Probably No − Definitely Yes		Definitely No − Probably Yes
	−2	Unsure − Definitely Yes	Probably No − Probably Yes	Definitely No − Unsure
	−1	Def. No − Prob. No	Prob No − Unsure Unsure − Prob. Yes	Prob. Yes − Def. Yes
	0	No Change		
	1	Def. Yes − Prob. Yes	Prob Yes − Unsure Unsure − Prob. No	Prob. No − Def. No
Gains	2	Definitely Yes − Unsure	Probably Yes − Probably No	Unsure − Definitely No
	3	Probably Yes − Definitely No		Definitely Yes − Probably No
	4	Definitely Yes − Definitely No		

[a] *XX* is a placeholder for the four environmental changes examined in our survey: beach width (*BW*), coral health (*CH*), marine life (*ML*) and seawater (*SW*).

Alternatively, we can analyze indicator variables representing different discrete response changes, such as changing the stated intention to return from "definitely yes" (DY) to "definitely not" (DN), or variables that represent groupings of response changes such as changing from "definitely yes" or "probably yes" (DPY) to "definitely not" or "probably not" (DPN):

$$Y_{DYDN} = 1, \text{ if } Y_1 = 5 \text{ and } Y_2 = 1; (Y_3 = 4)$$
$$Y_{DYDN} = 0, \text{ otherwise} \tag{2}$$

$$Y_{DPYDPN} = 1, \text{ if } Y_1 \geq 4 \text{ and } Y_2 \leq 2,$$
$$Y_{DPYDPN} = 0, \text{ otherwise} \tag{3}$$

The first step in our approach is to use correlation analysis and analysis of variance (ANOVA) to understand whether the values for *Change_Intention* are associated with the different values of environmental change presented in the survey. The second step is to use various forms of regression modeling to identify factors that are associated with changes in the stated intention to revisit.

A simple approach to modeling change in stated response is to use linear regression (an ordered logit specification would require conversion of *Change_Intention* to a value ranging from 0–8 rather than −4–4, and consequently the treatment of 0 as "no change" would not be preserved) to model *Change_Intention* as a function of the value of environmental change presented to the respondent, c_i, and individual characteristics, X_i:

$$Y_3 = (Y_2 - Y_1) = \beta_0 + \beta_1 c_i + \Sigma \ \beta X_i + \varepsilon_i \tag{4}$$

An important consideration is whether to include the initial willingness to return response (Y_1, *Plan_Return*) as a control variable in Equation (4). Including such a control ensures that any response changes are truly associated with other explanatory variables and not the result of differences in the initial response across groups in the sample. For example, if visitors from the U.S. are initially more likely to state a high probability of return (Y_1) relative to visitors from other countries, values of *Change_Intention* may be lower (more negative) for visitors from the U.S., all else being equal. Analysis of *Change_Intention* without the baseline response control might then lead to the erroneous conclusion that visitors from the U.S. are more sensitive to environmental change. After exploring the relationship between the initial willingness to return response (Y_1) and other respondent factors, we include Y_1 as a baseline control variable in modeling response change.

Another consideration for the model in Equation (4) is the assumption that the effect of environmental change, c_i, on *Change_Intention* can be captured by a single linear coefficient β_1. To examine whether the effect of environmental change on *Change_Intention* is non-linear, we can include a quadratic term, c_i^2, in Equation (4) or treat the alternative values of c_i as indicator variables.

As an alternative to modeling the change in intention to return as a continuous variable, binary response models such as logit regression can be used to understand factors that are associated with discrete changes such as those shown in Equation (2) or Equation (3). An individual i can be expected to change their stated revisit intention in response to an environmental change if their expected utility (satisfaction) U_i under the new revisit intention is higher than utility with the initial revisit intention:

$$U_{1i} (V_{1i} + \varepsilon_{1i}) \geq U_{0i} (V_{0i} + \varepsilon_{0i}) \tag{5}$$

where U_i represents the respondent's utility, which is comprised of an observable (deterministic) component V_i and an unobservable (stochastic) component ε_i.

The probability of changing responses to the willingness to return questions is therefore the probability that expected utility under the new conditions exceeds utility under the original (status quo) conditions. For example, the probability that an individual would change from "definitely will return" to "definitely will not return" is the probability that expected utility with certain intentions to not revisit (U_1) is higher than expected utility with certain intentions to revisit (U_0):

$$P_i (Y_{DYDN}) = P_i (Y_3 = -4) = P [U_{1i}(V_{1i} + \varepsilon_{1i}) \geq U_{0i} (V_{0i} + \varepsilon_{0i})] \tag{6}$$

To estimate probabilities such as Equation (6) using logit regression, we assume that utility is linear in the degree of environmental change presented to the respondent, c_i, and other respondent characteristics, X_i:

$$U_i (c_j, X_i) = \beta_0 + \beta_1 c_i + \Sigma \beta X_i + \varepsilon_i \tag{7}$$

For the logit specification, the probability shown in Equation (6) is then given by:

$$P_i (Y_{DYDN}) = \frac{\exp(\beta_0 + \beta_1 c_i + \Sigma \beta X_i)}{1 + \exp(\beta_0 + \beta_1 c_i + \Sigma \beta X_i)} \tag{8}$$

A consideration when examining response changes such as changing from "definitely will return" to "definitely will not return", is whether to limit the sample to respondents who were presented with scenarios involving negative environmental change. Respondents who were presented environmental improvements would not logically change from "will return" to "will not return", but will be coded as 0 in formulations of the dependent variable such as Equations (2) and (3), indicating that they did not make this change. Inclusion of these respondents may therefore confound the estimation of the coefficients on respondent characteristics. We explore Equation (8) using both the full sample and the subsample of respondents who faced scenarios involving negative environmental change.

3. Results

3.1. Summary Statistics

Visitors from the United States and Canada were over-represented in the sample relative to annual arrivals from Barbados' main source markets in 2015 (34% and 18% in our sample vs. 23% and 13% annual arrivals in 2015). Visitors from the U.K., the Caribbean and other European countries were under-represented (approximately 34%, 8% and 3% in our sample vs. 35%, 16% and 8% of annual 2015 arrivals). To accurately represent the preferences of the population of visitors by main source market, each observation was weighted using the ratio of the actual to sample percentage of visitors from each main market per year. All results presented below are based on the weighted sample.

Summary statistics are shown in Table 3. Notably, approximately 52 percent of the sample was visiting Barbados for the first time. Of those who had been to Barbados on a prior occasion, the average number of prior trips was approximately 6.5. Summary statistics for participation in coastal and marine recreation activities and ratings of coastal and marine quality are shown in Table 4. Respondents were heavily involved in coastal and marine recreation, with swimming, snorkeling and swimming with turtles being the most common activities. Respondents provided high ratings for all aspects of coastal and marine quality, with the quality of sand on beaches and the cleanliness and visibility of seawater receiving the highest average ratings and beach width receiving the lowest average rating.

Table 3. Descriptive statistics for respondent characteristics.

Variable	Definition	n	Mean	Std Dev	Minimum	Maximum
Environmental Change	Degree of environmental change presented in scenario	2548	−10.84	20.96	−50	25
First visit to Barbados	= 1 if respondent is had visited Barbados on a prior occasion = 0 otherwise	2538	0.52	0.49	0	1
Times to Barbados	Number of prior trips to Barbados	2476	3.02	6.31	0	100
US Resident	= 1 if the respondent is from the US = 0 otherwise	2548	0.24	0.42	0	1
UK Resident	= 1 if the respondent is from the U.K. = 0 otherwise	2548	0.41	0.48	0	1
Canadian Resident	= 1 if the respondent is from Canada = 0 otherwise	2548	0.14	0.34	0	1
Europe Resident	= 1 if the respondent is from other European country = 0 otherwise	2548	0.08	0.26	0	1
Caribbean Resident	= 1 if the respondent is from the Caribbean = 0 otherwise	2548	0.09	0.28	0	1
Higher Education	= 1 if the respondent had at least some college education = 0 otherwise	2548	0.56	0.49	0	1

Table 4. Summary statistics for participation in coastal and marine recreation and ratings of quality.

Variable	n	Mean	Std Dev	Minimum	Maximum
Participation in Recreational Activities					
Visited the Beach (%)	2459	0.92	0.27	0	1
Went Swimming (%)	2493	0.79	0.39	0	1
Snorkeled (%)	2493	0.42	0.48	0	1
Went Swimming with Turtles (%)	2493	0.37	0.47	0	1
Went Sailing (%)	2493	0.26	0.43	0	1
Went Jet Skiing (%)	2461	0.08	0.26	0	1
Went Power Boating (%)	2493	0.07	0.24	0	1
Went Scuba Diving (%)	2493	0.04	0.19	0	1
5-point Likert Scale Ratings of Coastal/Marine Quality 1 = "lowest quality", 5 = "highest quality"					
Quality of Sand on Beaches	2475	4.40	0.84	1	5
Cleanliness and Visibility of Seawater	2441	4.35	0.89	1	5
Natural Character of Beaches	2428	4.31	0.84	1	5
Ease of access to beaches	2445	4.31	0.92	1	5
Ease of getting in and out of the sea	2344	4.21	0.93	1	5
Cleanliness of Beaches	2484	4.19	0.93	1	5
Quality of coral reefs and marine life	1541	4.00	0.92	1	5
Width of Beaches	2449	3.97	0.97	1	5

Responses to the willingness to return questions are summarized in Table 5, which includes summary statistics for responses to the original willingness to return question, the follow-up questions involving environmental change and the *Change_Intention* variables. It is notable that changes in the quality of seawater induced significantly more changes in respondents' stated intention to return to Barbados than the other three environmental changes presented.

Table 5. Descriptive statistics for stated intention to return and *Change_Intention*.

	Variable	N	Mean	Std Dev	Minimum	Maximum
	Plan_Return	2578	4.33	0.80	1	5
	Definitely not	2578	0.005	0.07	0	1
Original stated	Probably not	2578	0.05	0.21	0	1
intention to return	Unsure	2578	0.06	0.23	0	1
	Probably	2578	0.41	0.48	0	1
	Definitely	2578	0.49	0.49	0	1
Intention to return	*Plan_Return_SW*	2577	2.94	1.43	1	5
with environmental	*Plan_Return_ML*	2561	3.68	1.67	1	5
change	*Plan_Return_CH*	2544	3.79	1.08	1	5
	Plan_Return_BW	2561	3.85	1.06	1	5
Change in intention to	*Change_Intention_SW*	2577	−1.39	1.45	−4	3
return with	*Change_Intention_ML*	2561	−0.65	1.07	−4	3
environmental change	*Change_Intention_CH*	2544	−0.53	0.95	−4	3
	Change_Intention_BW	2561	−0.47	0.89	−4	2

In Table 6, we show the percentage of respondents who changed their stated intentions to return to Barbados (i.e., the value of *Change_Intention* shown in Table 2 is not 0) by degree of environmental change. As expected, higher degrees of environmental change result in more respondents changing their stated intention to return. Two notable results are apparent from this Table. First, as illustrated in Table 5, significantly fewer respondents maintained their original willingness to return in response to changes in seawater quality relative to the other environmental changes presented. Even the smallest decline in water quality (5 percent higher chance of an infection) resulted in over 70 percent of respondents changing their response. Of the 352 respondents who were presented with the scenario involving a 50 percent reduction in water quality, 18 percent had the most severe change in intentions to revisit, changing their response from "definitely will return" to "definitely will not return" (*Change_Intention_SW* = −4). Another 44 percent changed from "definitely will return" to "probably will not return" or from "probably will return" to "definitely will not return" (*Change_Intention_SW* = −3). Hypothetical improvements in environmental quality did not influence the stated intention to return to Barbados for a large majority of respondents.

Table 6. Percentage of respondents who changed stated intention to return to Barbados, by severity of environmental change.

Environmental Change	Cleanliness & Visibility of Seawater (*SW*)	Quality of Marine Life (*ML*)	Coral Health (*CH*)	Beach Width (*BW*)
−50	90.96	66.02	60.22	61.50
−25	83.20	58.53	53.48	48.31
−10	76.87	46.45	41.65	35.60
−5	71.85	44.21	40.36	31.36
5	6.77	7.52	6.77	6.02
10	9.57	12.06	10.28	6.38
25	11.61	12.36	12.36	10.11

3.2. General Linear Models of Change_Intention Score

Correlation coefficients between the change intention variables (Table 2) and the degree of environmental change suggest significant positive correlations between these measures (Pearson correlation coefficients are 0.57, 0.41, 0.41 and 0.39 for changes in seawater quality, beach width, marine life and coral health respectively. All correlation coefficients are statistically significant at the $\alpha = 0.01$ level). Higher (lower) values of environmental change (which ranges from −50 to +25) are associated with higher (lower) values of *Change_Intention* (which ranges from −4 to +4). t-tests show that all values of environmental change resulted in statistically significant changes in responses for all environmental quality measures (all mean values of *Change_Intention* were statistically different than zero at the $\alpha = 0.01$ level for each level of environmental change). These results suggest that respondents changed their stated willingness to return in response to the hypothetical scenarios in the expected direction.

Analysis of variance (ANOVA) for differences in the *Change_Intention* variable for each pair of values of environmental change revealed no significant differences in *Change_Intention* for any pair of positive environmental changes. In other words, while environmental improvement scenarios caused respondents to change their stated intention to return in the expected (more favorable) direction, respondents did not react differently to 5 percent, 10 percent or 25 percent improvements in any of the environmental changes. Furthermore, we find no statistically significant differences in *Change_Intention* between 5 and 10 percent declines in beach width, coral health or the quality of marine life, or between 25 and 50 percent declines in coral health. Highly significant differences in *Change_Intention* were found for all other pairs of environmental change.

Numerous combinations of explanatory variables were tested in a GLM regression framework for *Change_Intention*. The environmental change variable was included as a continuous variable, a quadratic variable and using six indicator variables. Results shown in Table 7 are representative and include variables of policy interest (e.g., degree of environmental change, whether the respondent is a first-time visitor, respondent country of origin) and those that were found to be robust to specification. Because of the potential for biased estimates due to multicollinearity between the extensive and intensive margins of past visitation (i.e., the indicator variable for first time visitor and the number of prior visits), we examine model specifications that include both measures together and separately.

Table 7. General linear regression coefficient estimates. Dependent variable is *Change_Intention_XX*.

Parameter	Seawater	Beach Width	Coral Health	Marine Life
Intercept	2.65 ***	1.05 ***	1.46 ***	1.59 ***
Plan_Return	−0.50 ***	−0.24 ***	−0.31 ***	−0.37 ***
envchange −50	−2.56 ***	−1.15 ***	−1.24 ***	−1.50 ***
envchange −25	−2.12 ***	−0.85 ***	−1.04 ***	−1.22 ***
envchange −10	−1.79 ***	−0.58 ***	−0.87 ***	−1.10 ***
envchange −5	−1.61 ***	−0.51 ***	−0.74 ***	−0.90 ***
envchange +5	0.03	0.02	−0.06	−0.07
envchange +10	0.05	−0.01	0.02	0.04
Quality Rating [a]	−0.001	0.08 ***	0.06 **	0.06 **
First Visit to Barbados	−0.02	−0.04	−0.15 ***	−0.13 **
Previous trips to Barbados	0.01 +	0.002	0.001	0.01 *
US Resident	−0.22 ***	−0.18 ***	−0.09	0.01
UK Resident	−0.12 *	−0.04	−0.06	0.06
Canada Resident	−0.10	−0.09	−0.07	0.07
Caribbean Resident	−0.16	−0.18 **	−0.13	−0.03
Went Swimming	0.02	−0.10 **	0.03	0.03
Went to the Beach	−0.16 *	0.04	0.03	0.07
Snorkeled	−0.11 **	−0.02	−0.06	−0.11 **
Higher Education	−0.21 ***	−0.12 ***	−0.13 ***	−0.12 **
n	2348	2344	1480	1483
R^2	0.50	0.249	0.280	0.308

+, *, **, and *** indicate statistical significance at the 15%, 10%, 5% and 1% levels respectively. Standard errors not shown to save space. [a] Quality ratings are 5-point Likert scale ratings of cleanliness and visibility of seawater, beach width and the quality of corals and marine life.

In Table 7, positive (negative) coefficients indicate that higher values of the corresponding independent variable are associated with higher (lower) values of *Change_Intention*. For example, the negative coefficient on *Plan_Return* suggests that respondents who originally stated a higher probability of returning to Barbados were more likely to change in the negative direction (i.e., suggest a lower likelihood of return under conditions of environmental change).

Respondents who were presented with scenarios involving environmental losses were more likely to change their intention to return in the negative direction. On average, a higher magnitude of any form of environmental loss induces higher change in response, signaling that tourists are sensitive to declines in coastal and marine quality. In contrast, environmental improvement scenarios do not appear to induce significant change in stated intention to return. When coded as a continuous variable, coefficients on the environmental change variable were consistently positive and highly statistically significant, consistent with the results in Table 7. Including a quadratic term for environmental change in the specification (independently or in combination with the linear term) confirms that this relationship is nonlinear.

Tourists' nationality appears to explain differences in the responsiveness towards environmental degradation. Residents of the U.S. and U.K. appear more sensitive to changes in water quality, while U.S. and Caribbean visitors appear more responsive to reductions in beach width. We find no apparent differences in sensitivity to changes in coral health or marine life across nationalities. Respondents who provided higher Likert scale ratings of beach width, coral health or marine life are less likely to change their stated intention to return in response to changes in those aspects of environmental quality, implying that environmental losses are less of a deterrent for return visitation for tourists who are satisfied with these environmental conditions.

In terms of destination loyalty, first time visitors appear more sensitive to changes in coral health and marine life than repeat visitors, suggesting that repeat visitors are less likely to change their stated intentions to return following changes in these aspects of environmental quality. We also find mild evidence that respondents with more previous trips to Barbados are less sensitive to changes in seawater quality. That is, all else being equal, degradations in seawater quality are likely to have a smaller impact on return decisions for frequent visitors to Barbados.

All recreation activities shown in Table 4 were included as covariates, independently and in combination. Only swimming, going to the beach and snorkeling were found to be significantly associated with *Change_Intention*. It is notable that snorkelers appear to be more sensitive to changes in sea water quality and the quality of marine life than non-snorkelers. Finally, we find that respondents with higher education are more sensitive to all environmental changes than those without higher education.

To illustrate the differential impact of environmental change scenarios on respondents' stated intention to return, we plot fitted values of *Change_Intention* across all levels of environmental change using the coefficient estimates shown in Table 7. We calculate fitted values for a first-time visitor from the U.S. with a college education, who went to the beach, went swimming and snorkeling, and provided the highest ratings of environmental quality. The shapes of the functions in Figure 1 illustrate the nonlinear relationship between degree of environmental change and the magnitude of response in stated intention to return.

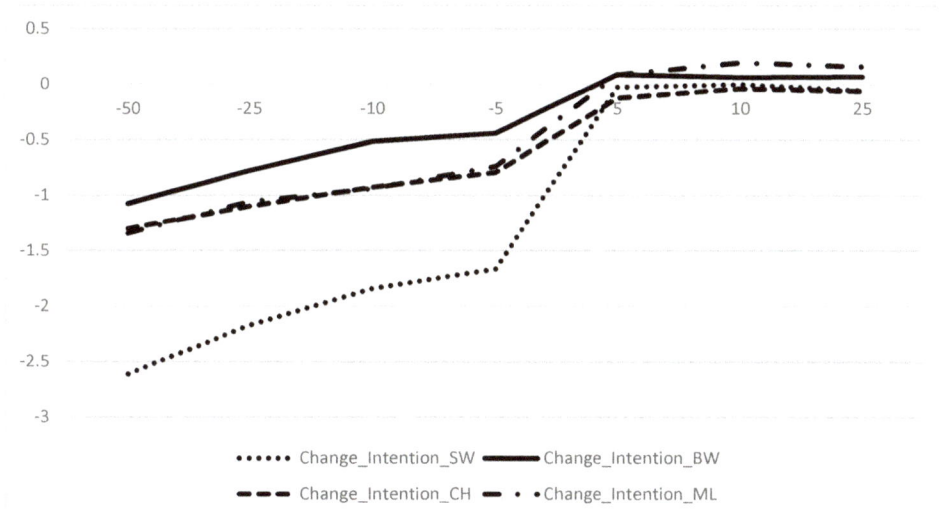

Figure 1. Predicted value of *Change_Intention* for different levels of environmental change. The value of *Change_Intention* measured on Y-axis. Degree of environmental change measured on X-axis.

3.3. Logit Models of Intention to Return

We use logit regression to identify the factors that are associated with discrete changes in stated intention to return. As shown in Table 2, above, respondents could change their stated intention to return in 20 distinct ways in response to hypothetical changes in environmental quality (10 for improvements in quality and 10 for declines). Because environmental improvements have little impact on stated intention to return, we restrict the sample to respondents who viewed scenarios involving reductions in quality.

Table 8 shows the results of logit regression models for changing responses from "definitely" or "probably" will return to "definitely" or "probably" will not return (i.e., the dependent variable is equal to one if the respondent changed their stated intention from "definitely" or "probably" will return to "definitely" or "probably" will not return, and equal to zero otherwise). Numerous model specifications and combinations of covariates were estimated. The results presented in Table 8 are representative. We also examined less severe changes, such as changing from "definitely will return" to "definitely" or "probably" will not return and found results very similar to those presented in Table 8. Limited responses for the most severe unfavorable change ("definitely will return" to "definitely will not return") inhibited maximum likelihood estimation.

Table 8. Logit regression coefficient estimates. Dependent variable = 1 if respondent changed response from "definitely" or "probably" will return to "definitely" or "probably" will not return.

Parameter	Seawater	Beach Width	Coral Health	Marine Life
Intercept	−4.23 ***	−3.51 ***	−3.50 ***	−3.09 ***
Plan_Return	0.75 ***	0.24 **	0.35 ***	0.47 ***
envchange −50	1.42 ***	1.45 ***	1.22 ***	1.19 ***
envchange −25	0.83 ***	0.74 ***	0.77 ***	0.77 ***
envchange −10	0.36 **	0.17	0.60 **	0.65 ***
Quality Rating	−0.05	−0.29 ***	−0.19 **	−0.14 *
First Visit to Barbados	0.18	0.39 **	0.80 ***	0.52 ***
Previous trips to Barbados	−0.02 **	0.0005	0.01	−0.02
US Resident	0.39 **	0.88 ***	−0.14	−0.34
UK Resident	0.37 **	0.49 *	−0.13	−0.41 *
Canada Resident	0.42 **	0.68 **	0.13	−0.25
Caribbean Resident	0.46 *	0.72 *	0.12	−0.08
Went Swimming	−0.16	0.52 **	−0.17	−0.16
Went to the Beach	0.41 *	−0.11	−0.03	−0.35
Snorkeled	0.36 ***	0.01	0.22	0.36 **
Higher Education	0.35 ***	0.17	0.24	0.17
n	1723	1722	1070	1073
% of Respondents who changed	60.5	16.1	19.1	25.6
AIC	2057.57	1313.73	961.17	1125.27
2 Log L	2025.57	1281.73	929.17	1093.27

[+], *, **, and *** indicate statistical significance at the 15%, 10%, 5% and 1% levels, respectively. Standard errors not shown to save space. [a] Quality ratings are 5-point Likert scale ratings of cleanliness and visibility of seawater, beach width and the quality of corals and marine life.

In these models, positive (negative) and statistically significant coefficients indicate that higher (lower) values of the variable are associated with a higher probability of changing the stated intention to return. For example, the coefficients on 50 and 25 percent environmental losses are consistently positive and highly significant. This indicates that these changes in environmental quality are associated with a higher probability of changing the stated intention to return to Barbados from "definitely" or "probably" will return to "definitely" or "probably" will not return. The positive coefficient on *Plan_Return* suggest that respondents who originally stated a higher probability of returning to Barbados were more likely to suggest a lower likelihood of return under conditions of environmental change.

As with *Change_Intention*, higher magnitudes of any form of environmental loss are associated with a higher probability that respondents change their stated intention to return. Again, we find nationality appears to explain differences in the response variable. Residents of the U.S., U.K. and Canada appear more sensitive to changes in water quality and beach width, and Caribbean residents appear to have similar but less significant tendencies to change response. We also find no significant differences in sensitivity to changes in coral health or marine life across main market points of origin, except for a mild reaction by U.K. residents to changes in marine life. Also supporting the above analysis of *Change_Intention*, environmental losses are less likely to affect return visitation for tourists who are satisfied with beach width, coral health or marine life, and snorkelers appear to be more sensitive to changes in sea water quality and the quality of marine life than non-snorkelers. We again find evidence of destination loyalty. First time visitors are more sensitive to changes in beach width, coral health and marine life than repeat visitors. Interestingly, first time visitors appear most sensitive to changes in coral health. Respondents with more previous trips to Barbados are again found to be less sensitive to changes in seawater quality.

4. Discussion

This research demonstrates that Barbados tourists' return visitation decisions are sensitive to declines in all aspects of coastal and marine quality. This general finding was expected, given that tourists' major attractions to Barbados are "sea, sand and sun" products and that tourists require value for their money. Of the four coastal and marine attributes examined, declines in seawater quality have the most significant impact on tourists' stated willingness to return, confirming the notion that perceived risks involving water-based recreation have important economic repercussions in tourism-dependent areas [10].

It is notable that even the highest levels of losses in beach width, coral health and marine life (i.e., 50% losses) do not produce as dramatic an impact on return visitation intentions as the smallest change in seawater quality (i.e., 5% higher probability of an infection). This finding is consistent with the results of Can and Alp, who find that tourists and residents are willing to pay more for water quality than improvements in marine life in Turkey [48], and Halkos and Matsiori, who find that improving the quality of bathing water is the most important determinant of willingness to pay for coastal zone improvement in Volos, Greece [14]. A possible explanation is that recreationists understand that seawater quality has a direct impact on their health, whereas attributes such as beach width, coral health and marine life are primarily associated with aesthetic enjoyment. It may also be the case that the quality of seawater has an impact on the aesthetic value and perception of reef quality and marine life; however, our data do not allow for testing of this hypothesis.

Another salient takeaway from this research is that scenarios involving environmental improvements do not result in statistically significant changes in tourists' stated intention to return. We can infer that tourists are averse to environmental losses, a result that may be an artifact of general loss aversion or an endowment effect. It may also be the case that respondents' expectations for environmental quality have largely been met. Ratings of coastal and marine quality were generally high, and those who found environmental quality to be favorable were less deterred by losses in quality. Satisfied with current conditions, further improvements do not induce significant changes in tourists' intentions to revisit. An important policy implication is that if the economic costs of improving environmental quality are prohibitive, maintaining status quo conditions might be enough to support continued tourism demand. It is important to note that this result stands in contrast to numerous monetary valuation studies that show positive WTP for improvements in coastal and marine quality [14,21–23,43], and illustrates the importance of understanding both price and quantity aspects of the demand relationship. It seems clear from the literature that people are WTP higher prices for trips with improved environmental quality, but it may be the case that such improvements would not induce a change in the number of trips demanded. This suggests that destinations can capture economic surplus from environmental improvements through pricing mechanisms, but not necessarily through increased visitation. Further research in this area appears warranted.

We find that first time visitors are more sensitive to changes in coral health and marine life than return visitors. It may be the case that repeat visitors have established stronger ties to the community and tend to visit for reasons beyond coastal and marine quality. This inference aligns with the marketing notion that it is harder to attract a new visitor than retain an existing one [12,50,51]. Existing repeat tourists have developed destination loyalty and may be harder to sway away from a familiar or preferred destination [52]. It is, therefore, not surprising that increased frequency of visitation (number of prior visits) reduces visitors' sensitivity to some environmental losses.

Respondents with higher education appear to be more sensitive to all losses in environmental quality and are especially sensitive to declines in seawater quality. More educated tourists may be more aware of environmental issues and the impact of environmental degradation on their health, thus having a higher aversion to adverse environmental effects. Furthermore, more educated tourists may be more aware of or more able to afford alternative destinations where environmental conditions are relatively favorable [14]. Importantly, we find that tourists in our sample with higher educations are younger, have higher incomes, and spend more on their trips to Barbados (all differences are highly statistically significant ($\alpha < 0.01$)). Our results therefore imply that continued deterioration of coastal and marine quality in Barbados may shift the demographic profile of tourists toward individuals who are less educated, less likely to engage in marine recreation and have lower economic impact while on island. Given our findings regarding nationality and stated intention to return, it also seems likely that continued losses in coastal and marine quality will result in lower return visitation from the U.K. and the U.S., Barbados' two main source markets.

Our findings are consistent with other results in the literature. For example, of the four coastal and marine attributes examined in Schuhmann et al. [44], tourists' WTP is least sensitive to changes in beach width and are only willing to pay to avoid very narrow beaches. We find that tourists who are more satisfied with environmental attributes are less sensitive to declines environmental quality, supporting the results of Um et al., who find that perceived attractiveness serves as a strong predictor of intention to revisit [50]. Our results suggest that visitation decisions by respondents who had recreational contact with coastal and marine environments (particularly through snorkeling) were more sensitive to environmental change than those who did not engage in these activities. This is consistent with results found by Beharry-Borg and Scarpa [21], Kosenius [44], and Hynes et al. [49], who find that willingness to pay for marine quality is positively associated with recreational contact and supports the findings of numerous studies in the literature showing that underwater recreationists have strong preferences and willingness to pay for healthy reefs and diverse fish populations [21–23]. Importantly, poor water quality is expected to have a greater impact on snorkelers than on swimmers, as water quality affects health and has an impact on the visual aesthetic of the marine environment.

Limitations of this research include the possibility for biased responses due to the hypothetical nature of our environmental change scenarios and the difficulties associated with separating environmental reasons for changes in aggregate visitation from other factors that affect demand. Future research could attempt to follow-up with respondents regarding actual revisit behaviors to empirically validate stated intentions. Combining information on the time trend of coastal and marine quality with estimates of return visitation over time could serve to isolate the overall impact of environmental change.

5. Conclusions

Repeat visitation signals satisfaction with a destination's attributes and is an important component in maintaining predictable tourist benefits and economic stability. With a return visitation rate near 50 percent, repeat visitors are critically important to the tourism-based economy of Barbados. Our examination of tourists' stated intentions to return to Barbados under scenarios of environmental change suggests that return tourism demand is highly vulnerable to continued losses in coastal and marine quality. Our results confirm the results of monetary valuation studies that suggest WTP by visitors to "sun, sea and sand" destinations is highly sensitive to coastal and marine degradation. Results presented here complement the existing WTP literature by providing non-monetary estimates of the potential consequences of continued nearshore degradation.

Of interest is the fact that seawater quality, a key driver of the health of coral reef ecosystems (i.e., coral and associated marine life) and indirectly related to beach width (as the producer of sand), was the attribute that visitors cared most about. These findings help to clarify policy focus regarding management actions designed to prevent coastal and marine degradation. In terms of maximizing returns from tourism, it seems clear that the focus of policy should be on preventing further declines in seawater quality.

These results are timely for Barbados. First, recent reductions in seawater quality and associated health hazards from sewage issues have resulted in negative media coverage. Governments from Barbados' main source markets (U.K., U.S. and Canada) issued travel warnings against visiting Barbados due to health concerns associated with this pollution, pointing to the need to understand the potential effect on the economy. Our results suggest that if the sewage issues are not resolved, the impact on the Barbadian economy via return tourism could be quite significant. Results from WTP studies in Barbados and elsewhere in the Caribbean suggest that recovering the costs associated with improving nearshore water quality may be feasible through pricing mechanisms such as conservation fees or higher prices for lodging and recreation, but our results imply that there may not be economic gains created through increased visitation.

Our results also support the Barbados Government's new strategic direction to develop a strong blue economy as seen in the 2018 creation of a 'Ministry of Maritime Affairs and Blue Economy' and the recent commitments of the Barbados Government at the Sustainable Blue Economy Conference in Kenya, November 2018. At this global gathering, the Government of Barbados announced the intention to follow a 'Roof to reefs' model (requiring management of land-based activities) in support of building its Blue Economy. Our results provide strong justification for this need to manage land-based activities to prevent further deterioration of coastal water. Efforts to restore reef ecosystems and mitigate coastal erosion are likely to result in lower returns if not coupled with management actions designed to prevent additional losses in seawater quality.

Beyond the importance for Barbados, this research highlights a relatively unexplored avenue for understanding the tradeoffs between environmental quality and the economic returns from tourism. The examination of how changes in environmental quality might affect people's behavior can provide insights that complement and extend the information gleaned from willingness to pay studies. First, understanding behavioral changes such as tourists' willingness to return to a destination can help bring the importance of natural resources and ecosystem services into the policy arena by highlighting the impact on tourism demand—a tangible and easily understood metric in locations where tourism is the principle economic driver. Furthermore, asking people to respond to behavioral questions avoids the difficulty of respondents having no prior experience with the monetary transactions proposed in WTP questions, which may lead to the expression of unreliable "constructive preferences" [53]. Estimates of behavioral change also avoid the moral, ethical and reductionist concerns associated with "commodifying" nature in monetary terms [54] and may minimize hypothetical bias in the case of familiar private goods such as travel [55]. Finally, willingness to return may be a more sensible expression of value than willingness to pay in the context of tourism. If visitors are not presented with a genuine opportunity to contribute monetary resources to conservation initiatives at a destination, they may simply "vote with their feet" and travel elsewhere.

Despite these advantages, contingent behavior studies examining the sensitivity of tourism demand to environmental quality are scarce in the literature. Given the relative ease of designing and analyzing willingness to return questions, valuation practitioners and tourist destinations that are dependent on natural resource quality would be advised to consider complementing existing assessments with hypothetical scenarios involving potential environmental change such as those used in this work.

Author Contributions: Conceptualization, P.S., R.S., R.W. and H.A.O.; Methodology, P.S., R.S., P.B.-S., and D.A.G.; Data Curation, P.S., R.S. and R.W.; Writing—Original Draft Preparation, P.S., R.S., R.W., P.B.-S., J.C., H.A.O. and D.A.G.; Writing—Review & Editing, P.S., R.S., P.B.-S., J.C., H.A.O. and D.A.G.; Supervision, P.S. and R.S.; Project Administration, P.S., R.S. and R.W.; Funding Acquisition, P.S., R.S. and R.W.

Funding: The APC was funded by the Cameron School of Business at the University of North Carolina Wilmington.

Acknowledgments: Support for this study was provided by the Caribbean Tourism Organization, the World Resources Institute and the Cameron School of Business at the University of North Carolina Wilmington. Additional support was provided by the Centre for Resource Management and Environmental Studies at the University of the West Indies—Cave Hill. Valuable consultation was provided by the Barbados Coastal Zone Management Unit and the Inter-American Development Bank. Mary J. Densmore and Jason A. Walsh provided significant administrative support.

Conflicts of Interest: The authors declare no conflict of interest.

References

1. Dharmaratne, G.S.; Brathwaite, A.E. Economic valuation of the coastline for tourism in Barbados. *J. Travel Res.* **1998**, *37*, 138–144. [CrossRef]
2. Burke, L.; Maidens, J. *Reefs at Risk in the Caribbean*; World Resources Institute: Washington, DC, USA, 2004; p. 80.
3. Bryant, R.L. Power, knowledge and political ecology in the third world: A review. *Prog. Geogr.* **1998**, *22*, 79–94. [CrossRef]
4. Burke, L.; Reytar, K.; Spalding, M.; Perry, A. *Reefs at Risk Revisited*; World Resources Institute: Washington, DC, USA, 2011; p. 144.
5. UNDESA. *World Urbanization Prospects*; the 2011 Revision; Population Division, Department of Economic and Social Affairs, United Nations Secretariat: New York, NY, USA, 2014.
6. Mycoo, M. Sustainable tourism using regulations, market mechanisms and green certification: A case study of Barbados. *J. Sustain. Tour.* **2006**, *14*, 489–511. [CrossRef]
7. Bell, P. Review paper: Eutrophication and coral reefs-some examples in the Great Barrier Reef lagoon. *Water Resour.* **1992**, *26*, 553–568. [CrossRef]
8. Hunte, W.; Wittenberg, M. Effects of eutrophication and sedimentation on juvenile corals. *Mar. Biol.* **1992**, *114*, 625–631. [CrossRef]
9. World Health Organization. *Guidelines for Safe Recreational Water Environments: Coastal and Fresh Waters*; World Health Organization: Geneva, Switzerland, 2003.
10. Pond, K. *Water Recreation and Disease: Plausibility of Associated Infections: Acute Effects, Sequelae, and Mortality*; World Health Organization: Geneva, Switzerland, 2005.
11. Loomis, J.B. An investigation into the reliability of intended visitation behavior. *Environ. Resour. Econ.* **1993**, *3*, 183–191. [CrossRef]
12. Mittal, V.; Kamakura, W.A. Satisfaction, repurchase intent, and repurchase behavior: Investigating the moderating effect of customer characteristics. *J. Mark. Res.* **2001**, *38*, 131–142. [CrossRef]
13. Grijalva, T.C.; Berrens, R.P.; Bohara, A.K.; Shaw, W.D. Testing the validity of contingent behavior trip responses. *Am. J. Agric. Econ.* **2002**, *84*, 401–414. [CrossRef]
14. Halkos, G.; Matsiori, S. Environmental attitudes and preferences for coastal zone improvements. *Econ. Anal. Pol.* **2018**, *58*, 153–166. [CrossRef]
15. García-Ayllón, S. New Strategies to Improve Co-Management in Enclosed Coastal Seas and Wetlands Subjected to Complex Environments: Socio-Economic Analysis Applied to an International Recovery Success Case Study after an Environmental Crisis. *Sustainability* **2019**, *11*, 1039. [CrossRef]
16. Casey, J.F.; Schuhmann, P.W. PACT or no PACT are tourists willing to contribute to the Protected Areas Conservation Trust in order to enhance marine resource conservation in Belize? *Mar. Policy* **2019**, *101*, 8–14. [CrossRef]

17. Schuhmann, P.W.; Skeete, R.; Waite, R.; Lorde, T.; Bangwayo-Skeete, P.; Oxenford, H.A.; David, G.; Moore, W.; Spencer, F. Visitors' willingness to pay marine conservation fees in Barbados. *Tour. Manag.* **2019**, *71*, 315–326. [CrossRef]

18. Trujillo, J.C.; Carrillo, B.; Charris, C.A.; Velilla, R.A. Coral reefs under threat in a Caribbean marine protected area: Assessing divers' willingness to pay toward conservation. *Mar. Policy* **2016**, *68*, 146–154. [CrossRef]

19. Pakalniete, K.; Aigars, J.; Czajkowski, M.; Strake, S.; Zawojska, E.; Hanley, N. Understanding the distribution of economic benefits from improving coastal and marine ecosystems. *Sci. Total Environ.* **2017**, *584*, 29–40. [CrossRef] [PubMed]

20. Christie, M.; Remoundou, K.; Siwicka, E.; Wainwright, W. Valuing marine and coastal ecosystem service benefits: Case study of St Vincent and the Grenadines' proposed marine protected areas. *Ecosyst. Serv.* **2015**, *11*, 115–127. [CrossRef]

21. Beharry-Borg, N.; Scarpa, R. Valuing quality changes in Caribbean coastal waters for heterogeneous beach visitors. *Ecol. Econ.* **2010**, *69*, 1124–1139. [CrossRef]

22. Gill, D.A.; Schuhmann, P.W.; Oxenford, H.A. Recreational diver preferences for reef fish attributes: Economic implications of future change. *Ecol. Econ.* **2015**, *111*, 48–57. [CrossRef]

23. Cazabon-Mannette, M.; Schuhmann, P.W.; Hailey, A.; Horrocks, J. Estimates of the non-market value of sea turtles in Tobago using stated preference techniques. *J. Environ. Manag.* **2017**, *192*, 281–291. [CrossRef]

24. Sorice, M.G.; Oh, C.-O.; Ditton, R.B. Managing Scuba Divers to Meet Ecological Goals for Coral Reef Conservation. *AMBIO* **2007**, *36*, 316–322. [CrossRef]

25. Rodrigues, L.C.; van den Bergh, J.C.; Loureiro, M.L.; Nunes, P.A.; Rossi, S. The Cost of Mediterranean Sea Warming and Acidification: A Choice Experiment among Scuba Divers at Medes Islands, Spain. *Environ. Res. Econ.* **2015**, *63*, 289–311. [CrossRef]

26. Shideler, G.S.; Pierce, B. Recreational diver willingness to pay for goliath grouper encounters during the months of their spawning aggregation off eastern Florida, USA. *Ocean Coast. Manag.* **2016**, *129*, 36–43. [CrossRef]

27. Alegre, J.; Cladera, M. Repeat visitation in mature sun and sand holiday destinations. *J. Travel Res.* **2006**, *44*, 288–297. [CrossRef]

28. World Travel and Tourism Council. *Travel and Tourism Economic Impact 2018*; World Travel and Tourism Council: London, UK, 2018.

29. BSS. Barbados Statistical Services Annual Report. Available online: https://www.barstats.gov.bb/ (accessed on 19 February 2019).

30. BTMI. *Annual Statistical Report*, 2017th ed.; Research Department, Barbados Tourism Marketing Inc.: Michael, Barbados, 2017.

31. Lewis, J.B. Evidence from aerial photography of structural loss of coral reefs at Barbados, West Indies. *Coral Reefs* **2002**, *21*, 49–56. [CrossRef]

32. Jackson, J.B.C.; Donovan, M.K.; Cramer, K.L.; Lam, V.V. (Eds.) *Status and Trends of Caribbean Coral Reefs: 1970–2012*; Global Coral Reef Monitoring Network, IUCN: Gland, Switzerland, 2014; p. 304.

33. Office of Research. *The Barbados Coral Reef Monitoring Programme: Changes in Coral Reef Communities on the West and South Coasts 2002–2012*; University of the West Indies: Wanstead, Barbados, 2014; p. 92.

34. Tomascik, T.; Sander, F. Effects of eutrophication on reef-building corals. I. Growth rate of the reef-building coral. *Mar. Biol.* **1985**, *87*, 143–155. [CrossRef]

35. Tomascik, T.; Sander, F. Effects of eutrophication on reef-building corals. II. Structure of scleractinian coral communities on fringing reefs, Barbados, West Indies. *Mar. Biol.* **1987**, *94*, 53–75. [CrossRef]

36. Tomascik, T.; Sander, F. Effects of eutrophication on reef-building corals. III. Reproduction of the reef-building coral. *Mar. Biol.* **1987**, *94*, 77–94. [CrossRef]

37. Snelgrove, P.V.R.; Lewis, J.B. Response of a coral-associated crustacean community to eutrophication. *Mar. Biol.* **1989**, *101*, 249–257. [CrossRef]

38. Marubinni, F.; Davies, P.S. Nitrate increases zooxanthellae population density and reduces skeletogenesis in corals. *Mar. Biol.* **1996**, *127*, 319–328. [CrossRef]

39. Holmes, K. Effects of eutrophication on bioeroding sponge communities with the description of new West Indian sponges, Cliona spp. (Porifera: Hadromerida: Clionidae). *Invertebr. Biol.* **2000**, *119*, 125–138. [CrossRef]

40. DeGeorges, A.; Goreau, T.J.; Reilly, B. Land-sourced pollution with an emphasis on domestic sewage: Lessons from the Caribbean and implications for coastal development on Indian Ocean and Pacific coral reefs. *Sustainability* **2010**, *2*, 2919–2949. [CrossRef]

41. SEDU. *The Sustainable Economic Development Unit (2002). Environmental Management Insertion in Tourism Sector Policies in the Caribbean*; Final Report; Department of Economics, University of the West Indies, St. Augustine Campus: St. Augustine, Trinidad and Tobago.

42. Hanley, N.; Bell, D.; Alvarez-Farizo, B. Valuing the benefits of coastal water quality improvements using contingent and real behaviour. *Environ. Res. Econ.* **2003**, *24*, 273–285. [CrossRef]

43. Schuhmann, P.W.; Bass, B.E.; Casey, J.F.; Gill, D.A. Visitor preferences and willingness to pay for coastal attributes in Barbados. *Ocean Coast. Manag.* **2016**, *134*, 240–250. [CrossRef]

44. Kosenius, A.K. Heterogeneous preferences for water quality attributes: The case of eutrophication in the Gulf of Finland, the Baltic Sea. *Ecol. Econ.* **2010**, *69*, 528–538. [CrossRef]

45. MacDonald, D.H.; Ardeshiri, A.; Rose, J.M.; Russell, B.D.; Connell, S.D. Valuing coastal water quality: Adelaide, South Australia metropolitan area. *Mar. Policy* **2015**, *52*, 116–124. [CrossRef]

46. Eggert, H.; Olsson, B. Valuing multi-attribute marine water quality. *Mar. Policy* **2009**, *33*, 201–206. [CrossRef]

47. Tait, P.; Baskaran, R.; Cullen, R.; Bicknell, K. Nonmarket valuation of water quality: Addressing spatially heterogeneous preferences using GIS and a random parameter logit model. *Ecol. Econ.* **2012**, *75*, 15–21. [CrossRef]

48. Can, Ö.; Alp, E. Valuation of environmental improvements in a specially protected marine area: A choice experiment approach in Göcek Bay, Turkey. *Sci. Total Environ.* **2012**, *439*, 291–298. [CrossRef] [PubMed]

49. Hynes, S.; Tinch, D.; Hanley, N. Valuing improvements to coastal waters using choice experiments: An application to revisions of the EU Bathing Waters Directive. *Mar. Policy* **2013**, *40*, 137–144. [CrossRef]

50. Um, S.; Chon, K.; Ro, Y. Antecedents of revisit intention. *Ann. Tour. Res.* **2006**, *33*, 1141–1158. [CrossRef]

51. Matute-Vallejo, J.; Bravo, R.; Pina, J.M. The influence of corporate social responsibility and price fairness on customer behaviour: Evidence from the financial sector. *Corp. Soc. Resp. Environ. Manag.* **2011**, *18*, 317–331. [CrossRef]

52. Baker, D.A.; Crompton, J.L. Quality, satisfaction and behavioral intentions. *Ann. Tour. Res.* **2000**, *27*, 785–804. [CrossRef]

53. Ajzen, I.; Brown, T.C.; Rosenthal, L.H. Information bias in contingent valuation: Effects of personal relevance, quality of information, and motivational orientation. *J. Environ. Econ. Manag.* **1996**, *30*, 43–57. [CrossRef]

54. Neuteleers, S.; Engelen, B. Talking money: How market-based valuation can undermine environmental protection. *Ecol. Econ.* **2015**, *117*, 253–260. [CrossRef]

55. Whitehead, J.C. Environmental risk and averting behavior: Predictive validity of jointly estimated revealed and stated behavior data. *Environ. Res. Econ.* **2005**, *32*, 301–316. [CrossRef]

Article

Using Internet Surveys to Estimate Visitors' Willingness to Pay for Coral Reef Conservation in the Kenting National Park, Taiwan

Nathaniel Maynard [1], Pierre-Alexandre Château [2,3,*], Lauriane Ribas-Deulofeu [3,4,5] and Je-Liang Liou [1,*]

[1]   Center for Green Economy, Chung-Hua Institution for Economic Research, Taipei 10672, Taiwan
[2]   Department of Marine Environment and Engineering, National Sun Yat-sen University, Kaohsiung 80424, Taiwan
[3]   Biodiversity Research Center, Academia Sinica, Taipei 11529, Taiwan
[4]   Biodiversity Program, Taiwan International Graduate Program, Academia Sinica & National Taiwan Normal University, Taipei 11529, Taiwan
[5]   Department of Life Science, National Taiwan Normal University, Taipei 10610, Taiwan; lauriane.ribas@gmail.com
*   Correspondence: pachateau@gmail.com (P.-A.C.); jlliou@cier.edu.tw (J.-L.L.)

Received: 30 April 2019; Accepted: 8 July 2019; Published: 9 July 2019

Abstract: Without appropriate conservation action, coral reefs globally continue to degrade, causing declines in economic value. Therefore, their local conservation and quantifying its benefits become increasingly important. However, accurately measuring these values remains expensive or complicated. Leveraging digital survey tools, an interdisciplinary on-line survey was created to estimate willingness to pay (WTP) for coral reef conservation using pictures and ecological data. Using the contingent valuation method we estimate current values as well as changes in value due to restoration or degradation for coral ecosystems in the Kenting National Park (KNP) in Taiwan. Results suggest that conserving degraded coral reef ecosystems leads to larger gains in value than healthier ones. Average WTP estimates a non-market economic value of 680 million US$ per year for the whole KNP marine area. Despite potential self-reporting bias and limits on sample size, these values appear consistent with similar studies and suggest future economic sampling strategies for KNP.

Keywords: contingent valuation method; internet survey; coral reefs valuation; non-market value

---

1. Introduction

The oceans provide trillions of dollars in economic and biodiversity values. It has been estimated that corals, mangroves and marine fisheries have a global asset value of US$ 6.9 trillion and that nearly three billion people rely on fish as a major source of animal protein [1]. However, due to pressures from pollution, climate change and overfishing, marine assets continuously decline [1,2].

Marine reserves have evolved to not only restrict fishing but also as resource management zones to reverse this trend [3]. Well managed marine resources provide sustained economic benefits over time primarily in the form of tourism, cultural, food and climate change mitigation value [4–6]. Proper marine protected areas management requires sustainable funding and adequate local capacity to increase biodiversity [7,8]. Unfortunately, global marine conservation remains inadequate, especially in the Asia-Pacific region [9,10].

1.1. Background on the Kenting National Park

Located in East Asia, Taiwan has tremendous marine and terrestrial biodiversity, especially in term of coral reef associated organisms [11,12]. Its reef associated biodiversity had put Taiwan among the worldwide ten most important marine hotspots of biodiversity [13]. The Kenting National Park (KNP) established in 1984 in the southern Hengchun Peninsula of Taiwan covers a terrestrial area of 181 km^2 and a marine area of 150 km^2 (Figure 1) and contains high levels of biodiversity, especially corals [14,15]. The KNP's impressive natural features and beaches draw millions of tourists per year (Figure 2). With a peak above eight million visitors in 2014, tourism has declined to just under four million in 2018.

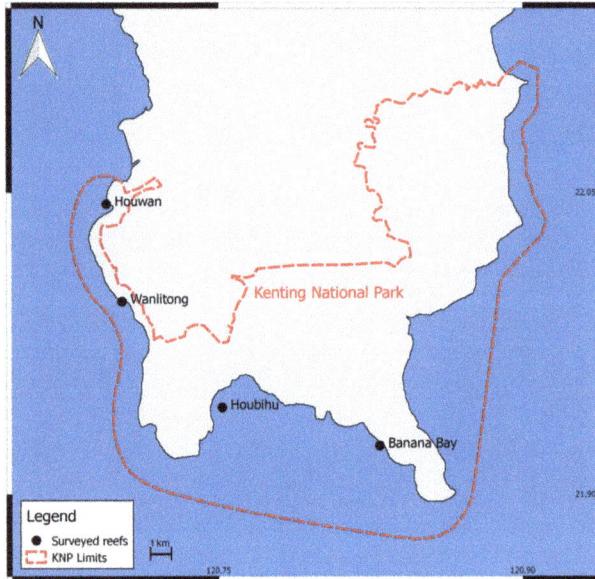

Figure 1. The Kenting National Park.

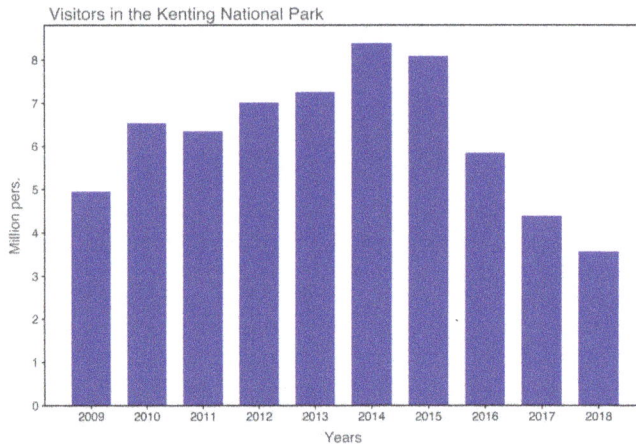

Figure 2. Visitors in the Kenting National Park. (https://www.ktnp.gov.tw/News.aspx?n=228F1362E45E0B89& sms=830F4DD99E91DBB7 (accessed on June 10 2019)).

Despite the recent decline in visitors, excessive nutrient pollution from tourism [16–18], persistent overfishing [19] and coastal development continue to degrade coral reefs [20–22]. Locally induced anthropogenic disturbances combined with typhoons and past bleaching events in Kenting are responsible for the loss of more than 50% of the coral coverage and an increase by almost three folds of macro-algae over the last three decades [22]. Rapid phase-shift from *Acropora* dominated state to sea-anemone *Condylactis* has been observed in Nanwan Bay [23]. A combination of bleaching events and typhoons followed by subsequent landslides and sewage overflows was suspected to be the cause of the rapid shift in the community as well as the local loss of biodiversity. Instead of the sudden regime shifts from destructive fishing practices commonly found in Asia-Pacific coral regions, the KNP reefs risk slowly yet irreversibly shifting to a lower biodiversity state with decreasing ecosystem services [24,25], which could also damage local livelihoods [26]. These corals may regenerate to a pre-disturbed state but are unlikely to unless conservation measures are enacted [15,23].

1.2. Coral Reef Valuation

Coral reef valuation started in the late 1980s focusing on coral reef degradation. By 2000, over 100 coral reef studies existed, and Brander et al. created the first meta evaluation based on recreational value [5]. Their analysis reveals that the worldwide average value of coral reef recreation is 184 US$ per person per visit. The median value, however, is 17 US$ per person per visit, showing that the distribution of values is skewed with a long tail of high values.

An evaluation of global ecosystem services found that coral reefs on average provided 352,249 INT$/ha/year, with most of this value coming from outside the market [27,28]. Another recent global study has found an average coral reef tourism value of 482,428 US$/km^2/year [6].

Several studies have estimated consumer surplus of national parks in Taiwan, although few have asked directly about coral [29,30]. Other studies have looked at the transportation cost of accessing Nanwan Beach, one of the most important coastal recreation areas in Kenting, but did not ask respondents specifically about coral reefs [31]. Others have asked about the impacts of oil spills [32]. Lastly, one study asked about the overall impact of climate change on coral for all of Taiwan [33].

All these previous studies look at different aspects of valuing coastal natural ecosystems in Taiwan. To improve conservation outcomes, how could one understand the benefits of conservation specifically for coral reefs in Taiwan updated with on-line sampling techniques? Would the public both adjacent to the park and far away support such measures? Due to high levels of tourism and challenges with park enforcement, the KNP is an ideal study site both in terms of potential information and need for updated economic valuation.

Natural resource economics has emerged to help policy makers and the wider community understand the economic benefits of conservation and how to fund such initiatives. Often, researchers use benefit transfers to quickly estimate values. A benefit transfer, where values from one natural resource valuation study are transferred from one site to another, allows for rapid estimates of ecosystem services [34]. While generally accepted to provide order of magnitude estimates, some have pointed out criticisms of benefit transfer regarding accuracy. Study quality and methodology can explain around 75% of variance between studies [35]. With such a variance in quality, cost and training become limiting factors for accuracy. Others have identified techniques to lower inaccuracy including adapting currency, matching transfers based on cultural similarity, habitat similarity and reef health [36]. Despite large amounts of tourism, valuation based on benefit transfers will not likely yield valuable results as Kenting has a relatively small (15,200 ha) fringing reef with very high tourism [37]. Spatially-based benefit-transfer estimations would therefore not yield accurate values.

Numerous studies have noted that visual criteria strongly influence willingness to pay (WTP) in corals [38], particularly water clarity [39] and fish abundance [40,41]. KNP has a diverse array of coral ecosystems across the peninsula spanning a wide mixture of soft and hard corals. Water clarity also changes due to a mix of anthropogenic and natural factors. Fish abundance and diversity depends on coral diversity and by association coral coverage [42].

Therefore, a rapid digital assessment of coral ecosystem value would help address key cost and accuracy concerns while also providing useful data for other sectors. This study aims to estimate visitors and residents WTP for coral reef conservation in the KNP. On this basis, a WTP function of coral coverage is fitted as a tool to estimate the economic value of different sites in southern Taiwan. The estimation results of this study provide a useful reference point for further usage of benefit transfer for evaluation of coral reef with diverse coverages.

2. Materials and Methods

2.1. Contingent Valuation Method

As a stated preference technique, contingent valuation (CV) has the potential to capture non-use and indirect values [43], which are crucial value components of coral reefs. CV surveys ask respondents to imagine a realistic scenario where they would spend money on a conservation activity in order to estimate the perceived value for that ecosystem or service [44].

The inspiration for asking questions digitally came from a study in Israel where the researchers used computer enhanced images to selectively remove coral, fish and other attributes in order to find values of each attribute [45]. Besides demographic information, we also wanted to collect visitation data, which can be useful for the tourism and public sector to better manage capacity. Other studies use hypothetical scenarios to estimate the value of restoring entire ecosystems and then apply those findings to larger regions [46]. Previous literature suggests that digital surveys do not have significantly different outcomes from in person surveys [47–49]. Moreover, an on-line questionnaire allows researchers to collect more responses than a typical survey, especially from users who do not frequently visit natural sites.

2.2. Survey Design

An initial pre-test was given to a group of 15 students at the College of Marine Sciences, National Sun Yat-sen University. Based on their responses, the survey was modified to improve clarity. The elicitation method in this phase was an open-ended format, and the results constituted the basis on which the payment card bidding levels were adjusted.

The survey was available on-line from May 2016 to mid-June 2016. The text was written in Mandarin Chinese on Google Forms and shared through email lists and Facebook posts with an emphasis on coral or dive groups and academic groups. Given funding limitations, we used a snowball approach to share the survey among various groups.

The survey begins with a short explanation on Kenting's biodiversity, threats to coral, as well as who the researchers were and what they plan to use the data for. The survey then collected demographic data including questions asking for gender, age, educational level, occupation, income, participation in environmental non-governmental organization (NGO) activities and place of residence. Respondents were then asked if they have been to the KNP before, how many times do they have visited in the past five years, how long did they stay and to indicate if they SCUBA dive or snorkel in the park.

After these questions about usage, they were asked valuation questions based on photographs of coral at different locations in the park. The question for willingness to pay was framed as a yearly payment into a "coral reef conservation fund" established by the government to maintain coral health at its current level. The site values were cumulative, as we asked the users at the start of the survey to pay for four sites. Below is the description text before the pictures:

"The KNP can be divided into four different regions based on ecosystem type (diversity and type of species) and health (quality and abundance). In Taiwan, the healthiest regions typically have coral covering 75% of the zone. However, with increasing tourism and pollution, coral coverage and diversity has decreased. To reverse this decline, the government will potentially set up a "Coral Protection Trust" (managed by an independent board) to enforce existing laws and prevent further degradation. In the next section, we will show you four pictures representing different regions and levels of coral coverage: Wanlitong (28% coral coverage), Houwan (32% coral coverage), Banana Bay (40% coral coverage) and Houbihu (50% coral coverage). We wish to know the amount you would be willing to pay for the conservation fund for each site to maintain its current state and prevent further degradation."

Participants were shown four separate pictures of different coral reefs around the KNP with varying levels of ecosystem quality (Figure 3). Each picture contained a short description and had a range of payments to select from (less than 3.2 US$/person/year to more than 64 US$/person/year, with increments of 3.2 US$/person/year).

2.3. Pictures Selection

Images were selected to best represent the ecosystem quality of the four zones. The main factors we looked at when assessing coral reef ecosystem health included: coral coverage, algae abundance and fish abundance. Using ecological field studies as a reference, the survey presented a sample of coral sites that formed a steady progression of coral coverage (Table 1). In the pre-test, respondents tended to value corals along a progression in quality regardless of actual coral quality; embracing this, the survey was modified to include descriptions of coral quality at each stage to mitigate this bias. Coral pictures were then ranked from least coverage to highest (Figure 3).

Table 1. Biological features of selected pictures.

Picture (Site)	Algae	Fish	Coral Coverage (%)	Coverage Level
1 (Wanlitong)	High	Low	28	4
2 (Houwan)	Medium	Low	32	3
3 (Banana Bay)	Low	Medium	40	2
4 (Houbihu)	Low	High	50	1

Figure 3. Pictures shown: (**a**) Wanlitong, (**b**) Houwan, (**c**) Banana Bay and (**d**) Houbihu.

Each picture included a short description of the area and any important features including location. Biological information is based on recent field data [15] and was summarized in very simple terms.

"Picture 1 (Figure 3a): Wanlitong (average coral reef coverage of 28%). This area is located between the marine ecological protection zone (Haishengyi) and the Haidi Park on the western coast of Kenting. It is characterized by a moderate coral coverage, a low number of fish and high biomass of seaweed.

Picture 2 (Figure 3b): Houwan (average coral reef coverage of 32%). This area is located on the western coast of Kenting, close to the homes of local residents and a large resort hotel. The environment is characterized by a moderate coral coverage, low quantities of fish and a moderate amount of seaweed.

Picture 3 (Figure 3c): Banana Bay (average coral reef coverage of 40%). This area is located in the ecological protection area (Haisheng III) on the east side of Nanwan Bay. There are only a few hotels in the vicinity, and the environmental characteristics are better than in Wanlitong. Banana Bay has a higher coral reef coverage, a moderate amount of fish and a small amount of seaweed.

Picture 4 (Figure 3d): Houbihu (average coral reef coverage of 50%). This area is currently a marine resource protection demonstration zone with rich coral reef ecology. Coral coverage is high, fish are diverse and abundant, and the amount of seaweed is relatively low. Houbihu is a popular area for scuba diving and snorkeling."

Before each picture respondents had to check a box to indicate whether or not they would be willing to pay for that specific site. If respondents indicated they would not be willing to pay, they would receive an additional question at the end asking them why they did not want to pay into the coral conservation fund. Several potential reasons were proposed, such as "It's the government's responsibility to protect coral" and "Corals have no value to me". Space was included to write in a personal reason that was not listed.

3. Results

Some degree of bias is unavoidable in WTP surveys especially with smaller sample sizes and with non-random sample groups. To minimize that interference, an ex-post data screening approach was used [50]: We used winsorization to limit the 5% extreme values. Self-selection bias for Internet surveys does not appear to be significantly stronger than that of in person surveys; as with any form of survey, there is some degree of self-selection [47].

We collected 296 responses of which 231 were considered appropriate for the survey. A total of 26 protest responses opted to fill out the survey but did not provide any WTP information, with the most common reason being "It's the government's responsibility to protect coral". Other reasons include: "Corals have no value to me", "I don't trust an organization to conserve coral reefs", "Local industries should take the responsibility of protecting corals", "Before contributing, I need to see for the effectiveness of conservation". The remaining 39 excluded responses either did not fill in the survey completely or improperly filled in answers.

3.1. Demographics

According to Table 2, collected demographic information was similar to average demographics in Taiwan although slightly biased towards the environment, higher salaries and with more men represented. One potential source of bias in the WTP values comes from 39% of respondents having participated in some form of environmental conservation activities, although that group could include non-profit membership to a hobby group such as diving.

Additionally, the sample had higher levels of educational level than the average in Taiwan. Previous studies have shown that educational attainment, income and environmental group membership increase WTP value [33]. The average age of the sample size was 33, which is slightly younger than the median age in Taiwan. The gender ratio was fairly split with males making up 53% (139) and females 47% (122).

The average wage (1344.79 US$/month) was likely driven up by higher incomes from professionals. While the most common response of 159.3 US$/month reflected the large amount of student responses (71). The median salary was 1115.1 US$/month.

Occupations fell primarily into service (48), student (71), industry (53), or public service/education (49). This means roughly 46% of respondents are in some way connected to academia or public service.

Table 2. Comparison between sample and population statistics (the gender ratio is defined as the ratio of men to women in 2016).

Comparison with Population	Monthly Income (US$/Month)	Schooling (Years)	Gender Ratio
Sample mean	1344.79	16.85	1.14
Population mean	1320.92 [1]	16.6 [2]	0.99 [3]
Relative difference	+1.8%	+1.5%	+15%

Note: 1: https://www.stat.gov.tw/np.asp?ctNode=522&mp=4 (accessed on May 22 2019); 2: https://www.stat.gov.tw/ct.asp?xItem=33332&CtNode=6020&mp=4 (accessed on May 22 2019); 3: https://www.ris.gov.tw/app/portal/346 (accessed on May 22 2019).

Next, we asked about visitation rates to KNP. We found the vast majority of respondents (239 i.e. 91.5%) had already been to KNP but only 190 (73%) within the last five years. Among them, only 98 (37.5%) saw coral during a snorkeling or scuba diving trip. This number is likely higher than the national average due to a large amount of responses from diving or environmental groups, although no such diving surveys could be found.

The average visitation rate in the past five years was 5.22 times, with a median of two and a mode of one visit. We also asked how many times participants went scuba diving with an average response of 17.95 and a median and mode of zero. This suggests that there are high frequency divers in the survey. Snorkeling was not as popular with the average reaching 2.36 times over five years, with zero and zero for median and mode respectively.

3.2. Willingness to Pay for the Four Pictures

The mean WTP (US$/person/year) for pictures 1 through 4 was equal to 16.0 ± 2.16, 18.55 ± 2.2, 20.46 ± 2.33, 25.15 ± 2.64, respectively. These values can be taken cumulatively to find a total WTP for the conservation of Kenting's corals of 80.160 ± 8.678 US$/person/year or 0.53 US$/km^2/person/year.

To test the equality of WTP across pictures, the nonparametric Kruskal–Wallis (KW) equality-of-population rank test [51] was employed (Table 3). According to the results of this test, almost all pairwise comparisons showed significant differences; only the WTP of pictures 2 and 3 were not distinguishable at the 10% level.

Table 3. KW tests for WTP across pictures (significant at *: 10%, **: 5%, ***: 1%).

KW Test	WTP1	WTP2	WTP3	WTP4
WTP1	-	3.143 *	6.145 **	24.55 ***
WTP2	3.143 *	-	0.498	11.664 ***
WTP3	6.145 **	0.498	-	7.759 ***
WTP4	24.55 ***	11.664 ***	7.759 ***	-

We organized the responses into a series of paired groups, according to the demographic information collected in the first section of the survey:

- People that once participated in an NGO activity vs. those who did not;
- Men vs. Women;
- Local people (Southern Taiwan) vs. tourists (Central and Northern Taiwan);
- Higher education (higher than college) vs. lower education (lower than college);
- Young people (less than 33 years old) vs older people (more than 33 years old).

We then used the KW test to assess whether the WTP of the paired groups were similarly distributed or not. The KW test did not find significant differences between the paired groups, but a small effect size was observable for NGO affiliation, residency and age. Table 4 provides the total WTP for each group and the effect size as measured by Cohen's d [52].

Table 4. Total WTP by groups (US$/person/year).

Groups		Sample Size	Total WTP	Effect Size (Cohen's d)
Activity with NGO	yes	90	85.792 ± 14.896	0.138
	no	141	76.566 ± 10.678	
Gender	M	119	79.115 ± 12.520	−0.032
	F	112	81.271 ± 12.172	
Locals	yes	65	74.258 ± 14.976	−0.123
	no	166	82.471 ± 10.629	
Education	High	114	80.740 ± 12.238	0.017
	Low	117	79.596 ± 12.482	
Age	<33	113	74.561 ± 10.808	−0.164
	>33	118	85.522 ± 13.559	
Overall	-	231	80.160 ± 8.678	-

3.3. Willingness to Pay as a Function of Coral Coverage

Using coral coverage as a proxy for ecosystem health at the four different sites, the WTP was expressed as a function of coral coverage (Figure 4). A concave utility function was fitted to the observed values, thereby providing a relation between WTP and coral coverage that could be used for the estimation of the WTP at other sites. This relation follows the law of diminishing marginal utility commonly observed for other economic goods. Visitors are willing to pay more for well protected sites than for degraded ones, but the rate of increase of their WTP decreases with coral coverage.

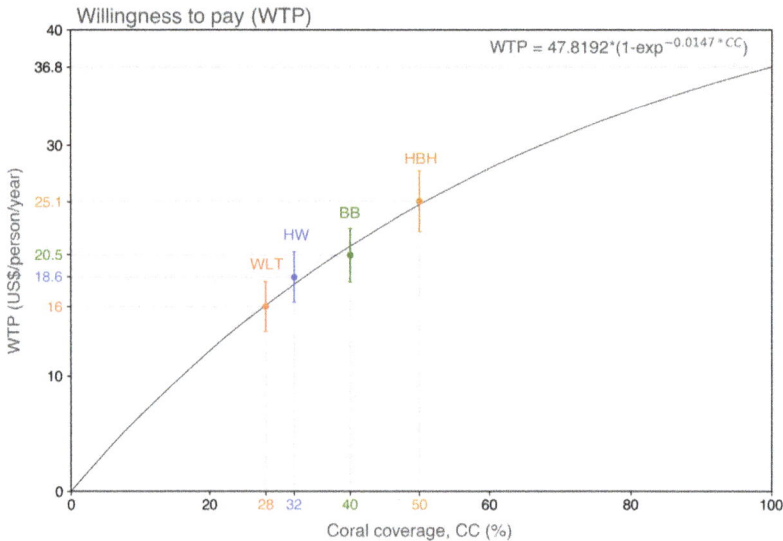

$$WTP = 47.8192 \cdot (1 - \exp^{-0.0147 \cdot CC})$$

Figure 4. Willingness to pay as a function of coral coverage (US$/person/year). WLT: Wanlitong, HW: Houwan, BB: Banana Bay and HBH: Houbihu.

The restoration of degraded ecosystems could provide larger gains in tourism value than an equivalent amelioration of healthier ones. Derived from this relation, Figure 5 shows the marginal cost of reef degradation as a function of coral coverage.

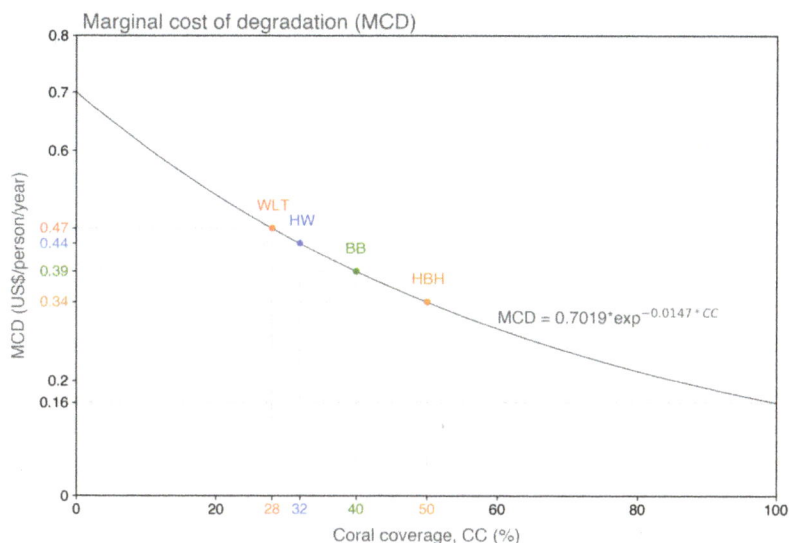

Figure 5. Marginal cost of degradation as a function of coral coverage (US$/person/year) WLT: Wanlitong, HW: Houwan, BB: Banana Bay and HBH: Houbihu.

Adopting such a marginal approach to conservation helps shifting the focus from highly covered reefs to reefs with a lower coverage but with higher potential increases in value. For example, a one point reduction in coral coverage is perceived as 38% more expensive in Wanlitong (0.47 US$/person/year) than it is in Houbihu (0.34 US$/person/year).

4. Discussion

This survey aimed to quickly and cheaply collect information on the value of the KNP coral reef ecosystems using pictures as a reference. Initial concerns included bias, response rate and picture clarity. Despite not paying for promotion, the survey quickly received a satisfying number of responses, with only few protest or improperly filled responses. Several findings are of particular relevance for local conservation.

First, individuals that were at least temporarily involved in an environmental NGO activity were willing to pay more (85.792 vs. 76.566 US$/person/year on average; Cohen's d = 13.8%) for coral conservation than those who were not, confirming previous results in Taiwan [33] and highlighting the importance of environmental education.

Second, age also influenced the WTP with people younger than 33 years old willing to pay less than people older than 33 (74.561 vs. 85.522 US$/person/year on average; Cohen's d = −16.4%). This is most likely due to differences of purchasing power between the younger respondents (mean income of 703.45 US$/month) and the older respondents (mean income of 1935.15 US$/month).

Third, there was another small size effect between locals (people residing nearby the KNP) and tourists (people residing elsewhere), with locals willing to pay less than tourists (74.258 vs. 82.471 US$/person/year on average; Cohen's d = −12.3%), reflecting the fact that for tourists corals are part of the reasons why they visit the KNP in the first place, so paying for their conservation is less a problem for them than for those who live nearby.

Finally, extrapolating average WTP to the 8.5 million households of Taiwan (https://census.dgbas. gov.tw/PHC2010/english/rehome.htm (accessed on June 14 2019)), the estimated contribution for conserving the KNP reaches 680 million US$/year or 4.5 million US$/km²/year. This estimation is an order of magnitude lower than the estimation of de Groot [27], which accounted for all ecosystem services, and an order of magnitude higher than Spalding [6], which focused on tourism. This suggests that our estimation may be somewhat optimistic. Indeed, the first economic valuation study in the KNP for an oil spill found an average WTP of 44.66 US$/person/year [32]. Another coral valuation study for all the reefs in Taiwan [33] found an average value of 35.75 US$/person/year. Our study reports values that are approximately twice higher, probably due to differences in sampling and methodology. Our questionnaire included pictures of coral reefs with specific biodiversity information. This may have positively influenced the respondents. The mode of survey was on-line, which may have made respondents more comfortable to record a higher value. On the other hand, the digital survey design, if properly marketed, can reach segments of society that do not frequently visit natural sites. Finally, numbers are likely higher due to the academic relationship of the majority of those surveyed and the participation in environmental group [53,54].

Larger sample sizes and additional surveys are required to determine more accurate values. Scope sensitivity has been studied extensively and continues to be debated within natural resource economics [55,56]. Moving economic values across cultures remains imprecise due to the spatial distribution of benefits and changes in price sensitivity. For the majority of studies, coral provides most value from fishing, wave blockage and tourism [27], but many corals in KNP do not provide significant shelter from waves or might have diminished or protected fisheries, while tourism may be much larger than in other reefs.

A larger goal of this research was to value biodiversity and resilience through revealed preference. By comparing the economic value of pictures with the real biological data, one could value a bundle of attributes as an ecosystem rather than the usual abstract category of "coral". This in turn would allow for better scenario analysis as we could model the changes in value due to ecosystem decline or restoration.

Expanding on this, another goal of this research was to value the marginal changes in ecosystem health. Does it make sense from an economic perspective to conserve a pristine area? Or would the benefits be greater if focusing on more degraded ecosystems? Our results show that respondents adjusted their WTP according to the given pictures and the short descriptions of environmental conditions that were provided. Their WTP for each picture differed significantly from one another, with the exception of pictures 2 and 3. This increasing WTP as we move towards healthier reefs allowed for the estimation of the WTP as a function of coral coverage (Figure 4).

Taking advantage of on-line tools can empower natural resource economics helping it reach more people and analyze a greater number of ecosystems. In the future, expanding the sample size to include a more representative sample of the Taiwanese population in terms of income, education level and environmental group membership would likely lead to a more accurate representation of value.

5. Conclusions

The KNP is a highly frequented area of international importance and high biodiversity, thus quantifying its benefits remains challenging. This study is, to our knowledge, the first in Taiwan to use Internet survey for coral reef valuation. Our results suggest that a total value of 680 million US$/year could theoretically be collected for conservation. We showed that areas of mid-range coral coverage could garner higher tourism value than highly covered sites. This implies that focusing conservation on marginal areas would lead to a better balance of conservation outcomes and tourism growth. While these results in no way suggest cutting funding from existing conservation, they instead point towards an opportunity for a higher value return on protecting the less well-preserved sites.

Author Contributions: Conceptualization, N.M., P.-A.C.; data curation, P.-A.C., J.-L.L.; investigation, N.M., L.R.-D.; methodology, N.M., P.-A.C., L.R.-D.; resources, J.-L.L.; supervision, J.-L.L.; writing—original draft, N.M.; writing—review and editing, P.-A.C., L.R.-D., J.-L.L.

Funding: During this study, N.M. was a recipient of the Fulbright Research Fellowship and P.-A.C. was the recipient of a postdoctoral fellowship from the Academia Sinica Sustainability Center (AS-104-SS-A03). L.R.D was the recipient of a Taiwan International Graduate Program scholarship (http://tigp.sinica.edu.tw/) and worked for the Academia Sinica Sustainability Project (AS-104-SS-A03).

Acknowledgments: The authors are thankful to the respondents who took the survey, especially those who shared it via their social networks. We also thank R.F. Wunderlich and two anonymous reviewers for their constructive comments.

Conflicts of Interest: The authors declare no conflict of interest.

References

1. Hoegh-Guldberg, O.; Global Change Institute; Boston Consulting Group. *Reviving the Ocean Economy: The Case for Action—2015*; World Wide Fund For Nature (WWF): Gland, Switzerland, 2015.

2. Cesar, A.H.; Burke, A.L.; Pet-Soede, A.L. *The Economics of Worldwide Coral Reef Degradation*; International Coral Reef Action Network, Cesar Environmental Economics Consulting (CEEC): Arnhem, The Netherlands, 2003.

3. Angulo-Valdés, J.A.; Hatcher, B.G. A new typology of benefits derived from marine protected areas. *Mar. Pol.* **2010**, *34*, 635–644. [CrossRef]

4. Moberg, F.; Folke, C. Ecological goods and services of coral reef ecosystems. *Ecol. Econ.* **1999**, *29*, 215–233. [CrossRef]

5. Brander, L.M.; Beukering, P.V.; Cesar, H.S.J. The recreational value of coral reefs: A meta-analysis. *Ecol. Econ.* **2007**, *63*, 209–218. [CrossRef]

6. Spalding, M.; Burke, L.; Wood, S.A.; Ashpole, J.; Hutchison, J.; zu Ermgassen, P. Mapping the global value and distribution of coral reef tourism. *Mar. Pol.* **2017**, *82*, 104–113. [CrossRef]

7. Waite, R.; Burkeand, L.; Gray, E.; van Beukering, P.; Brander, L.; McKenzie, E.; Pendleton, L.; Schuhmann, P.; Tompkins, E. *Coastal Capital: Ecosystem Valuation for Decision Making in the Caribbean*; World Resources Institute: Washington, DC, USA, 2014.

8. Anthony, K.R.N.; Marshall, P.A.; Abdulla, A.; Beeden, R.; Bergh, C.; Black, R.; Eakin, C.M.; Game, E.T.; Gooch, M.; Graham, N.A.J.; et al. Operationalizing resilience for adaptive coral reef management under global environmental change. *Glob. Chang. Biol.* **2015**, *21*, 48–61. [CrossRef] [PubMed]

9. Burke, L.; Selig, L.; Spalding, M. *Reefs at Risk in Southeast Asia*; World Resources Institute: Washington, DC, USA, 2002.

10. Burke, L.; Reytar, K.; Spalding, M.; Perry, A. *Reefs at Risk Revisited*; World Resources Institute: Washington, DC, USA, 2011.

11. Denis, V.; De Palmas, S.; Benzoni, F.; Chen, C.A. Extension of the known distribution and depth range of the scleractinian coral *Psammocora stellata*: First record from a Taiwanese mesophotic reef. *Mar. Biodivers.* **2015**, *45*, 619–620. [CrossRef]

12. Huang, D.; Licuanan, W.Y.; Hoeksema, B.W.; Chen, C.A.; Ang, P.O.; Huang, H.; Lane, D.J.W.; Vo, S.T.; Waheed, Z.; Affendi, Y.A.; et al. Extraordinary diversity of reef corals in the South China Sea. *Mar. Biodivers.* **2015**, *45*, 157–168. [CrossRef]

13. Allen, G.R. Conservation hotspots of biodiversity and endemism for Indo-Pacific coral reef fishes. *Aquat. Conserv. Mar. Freshw. Ecosyst.* **2008**, *18*, 541–556. [CrossRef]

14. Wu, T.-Y. Exploring Stakeholders' View of Management Effectiveness in Kenting National Park, Taiwan. Master's Thesis, Duke University, Durham, NC, USA, 2010. Available online: http://hdl.handle.net/10161/2153 (accessed on 9 July 2019).

15. Ribas-Deulofeu, L.; Denis, V.; De Palmas, S.; Kuo, C.-Y.; Hsieh, H.J.; Chen, C.A. Structure of Benthic Communities along the Taiwan Latitudinal Gradient. *PLoS ONE* **2016**, *11*, e0160601. [CrossRef]

16. Lin, H.-J.; Wu, C.-Y.; Kao, S.-J.; Kao, W.-Y.; Meng, P.-J. Mapping anthropogenic nitrogen through point sources in coral reefs using delta N-15 in macroalgae. *Mar. Ecol. Prog. Ser.* **2007**, *335*, 95–109. [CrossRef]

17. Liu, P.-J.; Meng, P.-J.; Liu, L.-L.; Wang, J.-T.; Leu, M.-Y. Impacts of human activities on coral reef ecosystems of southern Taiwan: A long-term study. *Mar. Pollut. Bull.* **2012**, *64*, 1129–1135. [CrossRef] [PubMed]

18. Chen, T.-H.; Chen, Y.-L.; Chen, C.-Y.; Liu, P.-J.; Cheng, J.-O.; Ko, F.-C. Assessment of ichthyotoxicity and anthropogenic contamination in the surface waters of Kenting National Park, Taiwan. *Environ. Monit. Assess.* **2015**, *187*, 1–16. [CrossRef] [PubMed]

19. Liu, P.-J.; Shao, K.-T.; Jan, R.-Q.; Fan, T.-Y.; Wong, S.-L.; Hwang, J.-S.; Chen, J.-P.; Chen, C.-C.; Lin, H.-J. A trophic model of fringing coral reefs in Nanwan Bay, southern Taiwan suggests overfishing. *Mar. Environ. Res.* **2009**, *68*, 106–117. [CrossRef] [PubMed]

20. Chang, Y.-C.; Hong, F.-W.; Lee, M.-T. A System Dynamic based DSS for sustainable coral reef management in Kenting coastal zone, Taiwan. *Ecol. Modell.* **2008**, *211*, 153–168. [CrossRef]

21. Meng, P.-J.; Lee, H.-J.; Wang, J.-T.; Chen, C.-C.; Lin, H.-J.; Tew, K.S.; Hsieh, W.-J. A long-term survey on anthropogenic impacts to the water quality of coral reefs, southern Taiwan. *Environ. Pollut.* **2008**, *156*, 67–75. [CrossRef] [PubMed]

22. Kuo, C.-Y.; Yuen, Y.S.; Meng, P.-J.; Ho, P.-H.; Wang, J.-T.; Liu, P.-J.; Chang, Y.-C.; Dai, C.-F.; Fan, T.-Y.; Lin, H.-J.; et al. Recurrent Disturbances and the Degradation of Hard Coral Communities in Taiwan. *PLoS ONE* **2012**, *7*, e44364. [CrossRef] [PubMed]

23. Chen, C.A.; Dai, C.-F. Local phase shift from Acropora-dominant to Condylactis-dominant community in the Tiao-Shi Reef, Kenting National Park, southern Taiwan. In *Coral Reefs*; Springer: Berlin, Germany, 2004; Volume 23, p. 508.

24. Tew, K.S.; Leu, M.-Y.; Wang, J.-T.; Chang, C.-M.; Chen, C.-C.; Meng, P.-J. A continuous, real-time water quality monitoring system for the coral reef ecosystems of Nanwan Bay, Southern Taiwan. *Mar. Pollut. Bull.* **2014**, *85*, 641–647. [CrossRef]

25. Keshavmurthy, S.; Meng, P.-J.; Wang, J.-T.; Kuo, C.-Y.; Yang, S.-Y.; Hsu, C.-M.; Gan, C.-H.; Dai, C.-F.; Chen, C.A. Can resistant coral-Symbiodinium associations enable coral communities to survive climate change? A study of a site exposed to long-term hot water input. *PeerJ* **2014**, *2*, e327. [CrossRef]

26. Lee, M.-T.; Wu, C.-C.; Ho, C.-H.; Liu, W.-H. Towards Marine Spatial Planning in Southern Taiwan. *Sustainability* **2014**, *6*, 8466–8484. [CrossRef]

27. De Groot, R.; Brander, L.; van der Ploeg, S.; Costanza, R.; Bernard, F.; Braat, L.; Christie, M.; Crossman, N.; Ghermandi, A.; Hein, L.; et al. Global estimates of the value of ecosystems and their services in monetary units. *Ecosyst. Serv.* **2012**, *1*, 50–61. [CrossRef]

28. Costanza, R.; de Groot, R.; Braat, L.; Kubiszewski, I.; Fioramonti, L.; Sutton, P.; Farber, S.; Grasso, M. Twenty years of ecosystem services: How far have we come and how far do we still need to go? *Ecosyst. Serv.* **2017**, *28*, 1–16. [CrossRef]

29. Yen, S.-C.; Chen, K.-H.; Wang, Y.; Wang, C.-P. Residents' attitudes toward reintroduced sika deer in Kenting National Park, Taiwan. *Wildl. Biol.* **2015**, *21*, 220–227. [CrossRef]

30. Dong, C.-M.; Lin, C.-C. Applying Count Models to Estimate the Tourism Demands and Recreation Benefits of Taijiang National Park. *Adv. Soc. Sci. Res. J.* **2019**, *6*, 256–273.

31. Dong, C.-M.; Lin, C.-C.; Lin, S.-P. Study on the Appraisal of Tourism Demands and Recreation Benefits for Nanwan Beach, Kenting, Taiwan. *Environments* **2018**, *5*, 97. [CrossRef]

32. Chen, C.-C.; Tew, K.S.; Ho, P.-H.; Ko, F.-C.; Hsieh, H.-Y.; Meng, P.-J. The impact of two oil spill events on the water quality along coastal area of Kenting National Park, southern Taiwan. *Mar. Pollut. Bull.* **2017**, *124*, 974–983. [CrossRef] [PubMed]

33. Tseng, W.W.-C.; Hsu, S.-H.; Chen, C.-C. Estimating the willingness to pay to protect coral reefs from potential damage caused by climate change: The evidence from Taiwan. *Mar. Pollut. Bull.* **2015**, *101*, 556–565. [CrossRef] [PubMed]

34. Lindhjem, H.; Tuan, T.H. Valuation of species and nature conservation in Asia and Oceania: A meta-analysis. *Environ. Econ. Policy Stud.* **2012**, *14*, 1–22. [CrossRef]

35. Liu, S.; Stern, D.I. *A Meta-Analysis of Contingent Valuation Studies in Coastal and Near-Shore Marine Ecosystems*; MPRA Paper 11720; University Library of Munich: München, Germany, 2008.

36. Londoño, L.M.; Johnston, R.J. Enhancing the reliability of benefit transfer over heterogeneous sites: A meta-analysis of international coral reef values. *Ecol. Econ.* **2012**, *78*, 80–89. [CrossRef]

37. Kung, T.A.; Lee, S.H.; Yang, T.C.; Wang, W.H. Survey of selected personal care products in surface water of coral reefs in Kenting National Park, Taiwan. *Sci. Total Environ.* **2018**, *635*, 1302–1307. [CrossRef]

38. Robles-Zavala, E.; Reynoso, A.G.C. The recreational value of coral reefs in the Mexican Pacific. *Ocean Coast. Manage.* **2018**, *157*, 1–8. [CrossRef]

39. Farr, M.; Stoeckl, N.; Esparon, M.; Larson, S.; Jarvis, D. The Importance of Water Clarity to Great Barrier Reef Tourists and Their Willingness to Pay to Improve it. *Tour. Econ.* **2016**, *22*, 331–352. [CrossRef]

40. Andersson, J.E. The recreational cost of coral bleaching—A stated and revealed preference study of international tourists. *Ecol. Econ.* **2007**, *62*, 704–715. [CrossRef]

41. Grafeld, S.; Oleson, K.; Barnes, M.; Peng, M.; Chan, C.; Weijerman, M. Divers' willingness to pay for improved coral reef conditions in Guam: An untapped source of funding for management and conservation? *Ecol. Econ.* **2016**, *128*, 202–213. [CrossRef]

42. Komyakova, V.; Munday, P.L.; Jones, G.P. Relative Importance of Coral Cover, Habitat Complexity and Diversity in Determining the Structure of Reef Fish Communities. *PLoS ONE* **2013**, *8*, e83178. [CrossRef] [PubMed]

43. Bartkowski, B.; Lienhoop, N.; Hansjürgens, B. Capturing the complexity of biodiversity: A critical review of economic valuation studies of biological diversity. *Ecol. Econ.* **2015**, *113*, 1–14. [CrossRef]

44. Carson, R.T. Contingent Valuation: A Practical Alternative When Prices Aren't Available. *J. Econ. Perspect.* **2012**, *26*, 27–42. [CrossRef]

45. Polak, O.; Shashar, N. Economic value of biological attributes of artificial coral reefs. *ICES J. Mar. Sci.* **2013**, *70*, 904–912. [CrossRef]

46. Bishop, R.C.; Chapman, D.J.; Kanninen, B.J.; Krosnick, J.A.; Leeworthy, B.; Meade, N.F. *Total Economic Value for Protecting and Restoring Hawaiian Coral Reef Ecosystems: Final Report*; NOAA Office of National Marine Sanctuaries, Office of Response and Restoration and Coral Reef Conservation Program; NOAA Technical Memorandum CRCP 16; NOAA: Silver Spring, MD, USA, 2011.

47. Lindhjem, H.; Navrud, S. Are Internet surveys an alternative to face-to-face interviews in contingent valuation? *Ecol. Econ.* **2011**, *70*, 1628–1637. [CrossRef]

48. Nielsen, J.S. Use of the Internet for willingness-to-pay surveys: A comparison of face-to-face and web-based interviews. *Resour. Energy Econ.* **2011**, *33*, 119–129. [CrossRef]

49. Marta-Pedroso, C.; Freitas, H.; Domingos, T. Testing for the survey mode effect on contingent valuation data quality: A case study of web based versus in-person interviews. *Ecol. Econ.* **2007**, *62*, 388–398. [CrossRef]

50. Loomis, J.B. 2013 WAEA Keynote Address: Strategies for Overcoming Hypothetical Bias in Stated Preference Surveys. *J. Agric. Resour. Econ.* **2014**, *39*, 34–46.

51. Kruskal, W.H.; Wallis, W.A. Use of Ranks in One-Criterion Variance Analysis. *J. Am. Stat. Assoc.* **1952**, *47*, 583–621. [CrossRef]

52. Cohen, J. *Statistical Power Analysis for the Behavioral Sciences*; Routledge: New York, NY, USA, 1988.

53. Spash, C.L. Non-Economic Motivation for Contingent Values: Rights and Attitudinal Beliefs in the Willingness to Pay for Environmental Improvements. *Land Econ.* **2006**, *82*, 602–622. [CrossRef]

54. Ahmed, M.; Umali, G.M.; Chong, C.K.; Rull, M.F.; Garcia, M.C. Valuing recreational and conservation benefits of coral reefs - The case of Bolinao, Philippines. *Ocean Coast. Manag.* **2007**, *50*, 103–118. [CrossRef]

55. Borzykowski, N.; Baranzini, A.; Maradan, D. Scope Effects in Contingent Valuation: Does the Assumed Statistical Distribution of WTP Matter? *Ecol. Econ.* **2018**, *144*, 319–329. [CrossRef]

56. Veisten, K.; Hoen, H.F.; Navrud, S.; Strand, J. Scope insensitivity in contingent valuation of complex environmental amenities. *J. Environ. Manag.* **2004**, *73*, 317–331. [CrossRef] [PubMed]

water

MDPI

Article

Applying Spatial Mapping of Remotely Sensed Data to Valuation of Coastal Ecosystem Services in the Gulf of Mexico

Valerie Seidel *, Daniel Dourte and Craig Diamond

The Balmoral Group, Winter Park, FL 32789, USA; ddourte@balmoralgroup.us (D.D.); cdiamond@balmoralgroup.us (C.D.)

* Correspondence: vseidel@balmoralgroup.us; Tel.: +01-407-629-2185

Received: 26 April 2019; Accepted: 30 May 2019; Published: 5 June 2019

Abstract: Spatial mapping of remote sensing data tends to be used less when valuing coastal ecosystem services than in other ecosystems. This research project aimed to understand obstacles to the use of remote sensing data in coastal ecosystem valuations, and to educate coastal stakeholders on potential remote sensing data sources and techniques. A workshop program identified important barriers to the adoption of remote sensing data: perceived gaps in spatial and temporal scale, uncertainty about confidence intervals and precision of remote sensing data, and linkages between coastal ecosystem services and values. Case studies that demonstrated the state of the science were used to show methods to overcome the barriers. The case studies demonstrate multiple approaches to valuation that have been used successfully in coastal projects, and validate that spatial mapping of remote sensing data may fill critical gaps, such as cost-effectively generating calibrated historical data.

Keywords: coastal ecosystems; remote sensing; ecosystem services valuation

1. Introduction

Ecosystem services valuation studies tend to be very local in scope, as summarized eloquently by Barbier (2012) [1]: the value of selected species impacted in the Greater Everglades restoration [2], the recreation value of coastal marshes in Saginaw Bay [3], etc. Policymakers frequently need to scale up to a broader geography for decision-making purposes. Spatial mapping of satellite data (hereafter, Earth Observations, or "EO" data), and other remotely sensed data, provides a convenient vehicle for extrapolating the right valuation data to a desired geographic or temporal scale. The research objective for this project was to identify barriers to the greater use of EO data in coastal ecosystem valuation, and potential solutions to improve the use of EO data.

Over the past two decades, remotely sensed data have been the standard bearer for many data sources underpinning ecosystem services assessments. Land use and land cover data products derived from aerial photography are common uses of remotely sensed data in valuations of social and environmental benefits from natural resources. Ecosystem services mapping presents its own challenges, as noted by Drakou (2015) [4] and others [5].

Coastal ecosystems, however, have been less well represented in the use of remotely sensed data. Coastal ecosystems do not fall easily into the land use nor marine categories for which many if not most ecosystem services assessments have been completed using remote sensing. A review of The Economics of Ecosystems and Biodiversity database ("TEEB") of ecosystem services values reveals that 312 of the 1310 entries address nine coastal ecosystems, 119 of which are focused on just four ecosystem services (food, climate, recreation, and extreme events) [6]. As Drakou [7] notes, ecosystems and their services may occur in locations separated by significant distances at the coast.

A 2018 review of coastal ecosystems references in TEEB finds that sea-grass beds are the least considered ecosystems, while tourism and recreation services are the most common ecosystem service considered [8]. Ecosystem services particularly important to coastal resource managers, like storm protection and erosion reduction, have not been as frequently valued in the literature due to the sophisticated hydrodynamic models required to estimate the impacts of different coastal ecosystems on flooding and land loss/gain [9]. Meanwhile, resource managers, whether in administering estuarine research and preservation, land management in coastal communities, or habitat restoration in coastal watersheds, spend precious financial and staff resources collecting data to estimate coastal ecosystem condition and to monitor trends in coastal ecosystems as part of ongoing project work.

Based on discussion with coastal managers of the types of data being collected, at least some of the data collection effort currently taking place could potentially be replaced with EO data sources. The research team proposed a series of workshops in Gulf Coast states to advance the use of Earth observations for valuing ecosystem services and to define practical means to incorporate these values into resource planning and decision making in the coastal environment.

Partners in the workshops would include public and private sector stakeholders that share the common goal of improving the resilience of coastal infrastructure and natural systems. Coastal infrastructure, including (constructed) living shorelines and natural barrier islands, dunes, shellfish beds, and coastal wetlands, have been identified generally as having significant potential to improve community resilience against storms and sea level rise, rebuild coastal ecosystems and increase the ecological benefits and services that contribute to coastal economies. However, specific economic measures of such benefits are either lacking or are insufficiently quantified to provide practical information to decision makers. This is particularly the case where such ecological services compete for weight or consideration against more broadly understood goods and services recognized by markets and the various economic measures historically used by decision makers.

The workshops would be designed specifically to evaluate key elements of implementing ecosystems services into the resource planning processes, such as the following:

- What data sources, including Earth observations, are needed to improve the approaches for valuing coastal ecosystem services relating to resilience?
- What valuation techniques best translate quantified ecosystems services defined for the Gulf Coast area?
- To what degree do resource managers currently use the values of ecosystems services in decision making?
- What institutional and informational barriers exist that constrain the application of ecosystem services within decision making related to planning and management of coastal resources?
- What steps need to be undertaken to accelerate the adoption of ecosystem valuation and elevate its impact on decision making?

The aim of the workshop series was to include guidance to NASA and other Federal agencies on the use of earth observations to inform ecosystems valuation within a regional resource—the Gulf of Mexico—with numerous interacting agencies, Federal and state.

2. Methods

Via a series of workshops with coastal resource managers, we set out to investigate the reasons for the under-utilization of EO data for coastal ecosystems valuation. The workshops were prefaced with a literature review, telephone interviews and surveys to understand the familiarity with EO data for coastal ecosystems, and barriers to its application. The workshop series was intended to be an exercise in iterative learning so that each session built on its predecessors in terms of constructing a knowledge base about the application of ecosystems valuation in theory and practice, as related to local and Federal agency resource planning and coordination.

Workshop Participants, Locations, Contents

Based on feedback from the telephone interviews conducted during the first six months of the project, workshops were scheduled for locations in each of the five Gulf of Mexico States. Venues were selected from public facilities used regularly by the National Estuarine Research Reserves (NERRs), National Estuary Programs (NEPs), and their stakeholders (like the Harte Research Institute at Texas A&M). Target attendees were identified through existing coastal information networks with the assistance of key stakeholders in each region. Resource managers from federal, state and local governments, coastal researchers, and not-for-profit groups focused on coastal conservation, science, and management were identified and invited to participate in an online survey, telephone interview, and workshop registration. Five workshops were delivered, one in each of the five Gulf Coast states over a 7-month period during the 18-month project.

Prior to each workshop, a custom survey was drafted to reflect the specific issues to each region and refine the specificity of issues toward the project objectives. The results of each survey were used to adjust the agenda to the user audience, and specifically to their identified level of familiarity with remote sensing tools, datasets and application to coastal ecosystem services.

Workshop content was prepared based on literature review, extensive interviews of experts and field practitioners, and steering committee members. A database of about 130 peer-reviewed publications or white papers related to spatial ecosystem services was assembled and utilized to guide workshop content development. Publications included those about research in remote sensing for modeling ecosystem services [10] and those about applications of ecosystem service information for management or planning improvements, particularly in coastal areas [11–13]. Additionally, numerous publicly available ecosystem service models were evaluated, and the most relevant EO data access and analysis portals were assembled for inclusion in the training portions of the workshops.

Evolving research of ecosystem services assessments has identified that beyond the challenges of logistics and scientific complexity, numerous social factors, including "management regimes, power relationships, skills, and values" [14], can influence approaches to ecosystem services. In response, participatory methodologies have emerged to capture the human aspects of interdependence and uncertainty in ecosystem services assessments [15,16]. One area of interest for the workshops was an understanding of the factors beyond scientific complexity that may influence current practice. In several of the workshops, a local expert was invited to review a local case study in ecosystem services, and describe the application of remotely sensed data to the region.

3. Workshop Results

Each workshop was near capacity and attracted between 25 and 35 attendees with varying levels of background in EO data, ecosystems services and valuation methods. Generally speaking, our research found that coastal managers and researchers in the Gulf States are not commonly employing EO data to value ecosystem services. Figure 1 reflects workshop participant responses regarding their pre-workshop use of ecosystem services quantities estimates.

Data Types used to Quantify Ecosystem Services

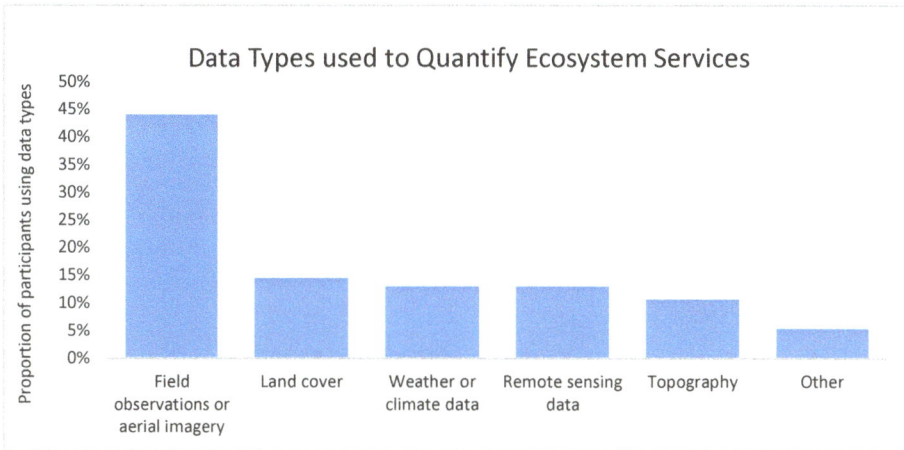

Figure 1. Workshop participant responses regarding current practice.

In a broader context, coastal managers tend not to attempt valuation of ecosystem services in general, relying instead on more traditional areal extent measures to quantify ecosystems (i.e., hectares of seagrass, acres of emergent marsh, etc.). Most practitioners are regularly collecting data on coastal ecosystems; some have collected data on ecosystem services; few have attempted to value ecosystem services, whether the value in question is economic or some other measure of societal benefit. Figure 2 summarizes the stated purpose for existing efforts to quantify ecosystem services, as reported by workshop participants.

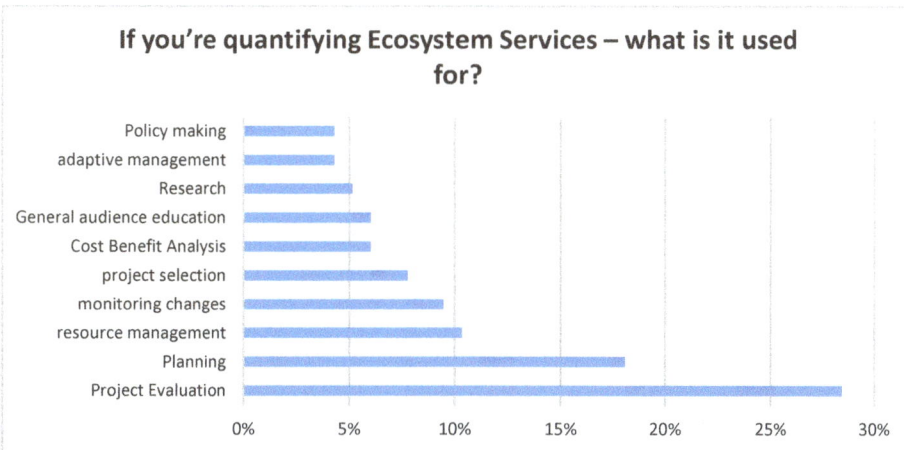

If you're quantifying Ecosystem Services – what is it used for?

Figure 2. Workshop participant responses regarding uses of ecosystem services quantities.

General awareness of EO data that could be productively used in coastal ecosystem valuation was low, but varied significantly between states. The critical finding, beyond awareness, relates more to the linkage between remote sensing data and locally collected data.

Interpolation of multiple scales of spatial and/or temporal resolution data was not routinely considered, and as a result practitioners tended to dismiss EO data for detailed analysis. Similarly, overlaying or merging other datasets that could create the finer level of information was generally dismissed. Using coastal or local case studies increased awareness of the process, and reduced resistance to lower resolution datasets. The workshops used specific case studies that improved the state of the science as examples, including those described in the next section.

While the experience of the practitioners attending the workshops was varied, there remained a clear interest in expanding the use of EO data for a range of management needs (within their constraints of time and priorities). The use of EO for ecosystems valuation was generally considered important as, in their view, valuation outcomes are often critical in discussions with policy and decision makers. The issue for many was ensuring a comfort level with valuation, quantifying uncertainty, and understanding practical applications. Figure 3 summarizes the responses to which ecosystems services matter most in coastal managers' regular work activities.

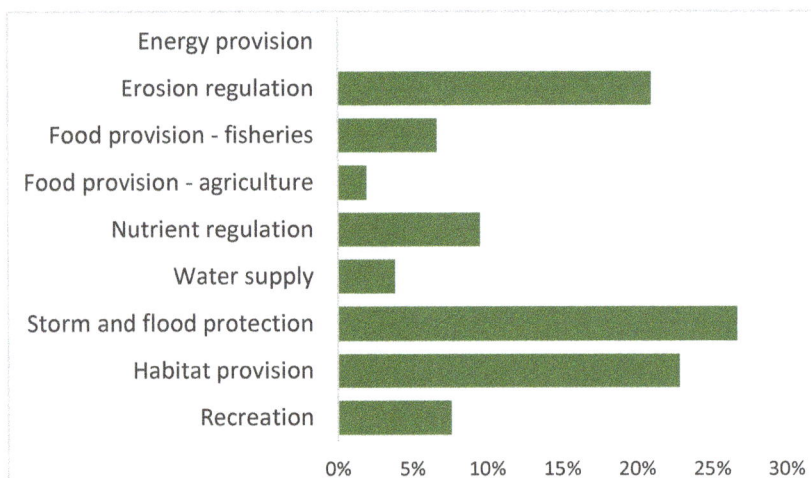

Figure 3. Workshop responses to which ecosystem services matter most in participants' work.

The case study examples in the next section demonstrate varied applications of earth observation data to improve coastal ecosystem management.

4. Case Studies in Using Earth Observations for Coastal Ecosystem Valuation

4.1. Evaluating Costs and Benefits of Coastal Resiliency Strategies

The State of Florida recognized that increasing coastal hazards would require consideration of new planning strategies for local governments. Required by statute to prepare "Comprehensive Plans", local governments establish rules regarding development, construction, and levels of municipal services that are obligated to local residents. The State contracted The Balmoral Group to perform a cost–benefit analysis of coastal resiliency strategies for consideration in Comprehensive Plan updates, including transfers of development rights, zoning changes, planned retreat, and armoring (sea walls). The State's interest was two-fold: (1) identifying the level of incentives required to motivate responsible development without a command regulation approach and (2) offering Comprehensive Plan language for counties to use in addressing development or redevelopment in hazardous coastal areas.

Florida's population and landscape are diverse, and two counties volunteered as pilots for the analysis—Martin County, which is Atlantic Ocean-facing, and home to a higher income, relatively less diverse population of about 160,000; and Okaloosa County, which resides on the Gulf of Mexico, houses the largest military base in the U.S., and partly because of the large base, includes more undeveloped land for its 200,000 residents. All properties seaward of the Coastal High Hazard Area were included in the analysis. Estimating realistic costs and benefits to identify thresholds at which incentives would change behavior required assessing conditions at the site level: economics, wave energy, environmental factors, and others. With thousands of public and private properties in play, remote sensing data were used to calculate specific values at the property level in a Geographic Information System, or GIS.

Both cost and benefit factors were assessed in GIS at the property level, including acquisition costs, construction costs for armoring or demolition, amenity values, avoided costs, and foregone revenues. Publicly available parcel data were obtained from property appraisers and overlaid with high resolution aerial photography, land use/land cover data for wetlands delineation and permanently preserved lands, and beach zones. Remote sensing data played a key role in efficiently assigning values. For example, armoring costs were estimated by linear foot depending on the wave energy with decision rules developed with coastal engineers; values ranged from $5250 to $10,800 per meter (in 2019 dollars). Ecological values for beach habitat were estimated based on public values for ecological protection (of habitat and sea turtle nesting sites) of $11/resident and applied evenly to waterfront parcels adjacent to critical beach areas [17]. Public values for recreational access to the beach were calculated using Willingness to Pay (WTP) values of $67–183/visit and beach visitor data; and values for wetland preservation, where wetlands may allow seagrass migration to occur, ranged from $37,000 to $79,000, with an inflation adjusted mean of $53,300/ha [18]. The resulting values contributed to a comprehensive evaluation of feasible coastal resiliency strategies; ultimately, the analysis found that all strategies were feasible in some areas under current conditions, but the results were place-specific and changed considerably under alternative future climate scenarios.

The Coastal Resiliency Strategies example provides a solid case of using remote sensing data to estimate values with spatially distinct variation across large areas and subjects. Gathering site-level detail through field work would have proven too expensive and time-consuming for an economic analysis of cost and feasibility. Combining engineering (wave energy, armoring costs) and scientific input (critical erosion areas, sea turtle nesting sites, and wetland delineation) with economic values, calculations across thousands of individual sites were completed rapidly and with precision. It is likely that this sort of analysis would have previously been addressed through county-level averages, potentially painting a very different picture of which strategies would be feasible for a given neighborhood, street or community. The values can readily be updated, using annually updated datasets.

4.2. Restore the Balance: TBEP Applies Remote Sensing Data to Understand Habitat Mosaic and Ecosystem Services

The Tampa Bay Estuary Program (TBEP) recognized the need for a habitat mosaic approach. Upstream freshwater wetlands were critical to certain estuarine species lifecycles, and diminishing populations of these species was causing economic consequences to tourism and recreational fishing. Understanding that different basins/hydrologic units in the watershed had changed in different ways, TBEP undertook analysis to assess trends from pre-urbanization to recent conditions.

For the region, the earliest data available dated to military flight aerial photography from the early 1950s. Digitized historical aerials, national wetland inventory data and national hydrography data [19] were combined in GIS to evaluate trends in wetland composition and to build Landscape Development Intensity (LDI) values for 1952 and 2007, based on the work of Brown and Vivas [20]. Changes in wetlands, by basin and type, were quantified, recognizing that a greater percentage loss of specific types of wetlands at key locations within a particular basin or hydrologic unit may mean substantial changes to ecosystem dynamics.

(Note: LDI values are expressed in units of emergy or solar emjoule (seJ/ha/year). In this case, the output of manual inspections of surrounding land use, wetland condition and other factors in Florida's Uniform Mitigation Assessment Method was compared to predictions by LDI.)

LDI values were estimated in GIS on a 10 m × 10 m grid. Conditional assessments were performed via site visits to assess condition and vulnerability, understand types of change, and calibrate LDI measurements. The conditional assessments considered location and landscape support, water environment, and community structure; the LDI predicted observed field conditions reasonably well, with an R^2 of 0.69. Based on the findings of 37 site visits, the LDI calculations were calibrated and completed for the remaining roughly 5200 square kilometer watershed at 0.2 ha accuracy. The data were then used to estimate coastal ecosystem services, including relative nutrient reduction and regulation, water supply, and flood attenuation. Figure 4 shows example maps reflecting spatial mapping of projected LDI based on future land use plans in the left panel, while the right panel shows wetland connectivity.

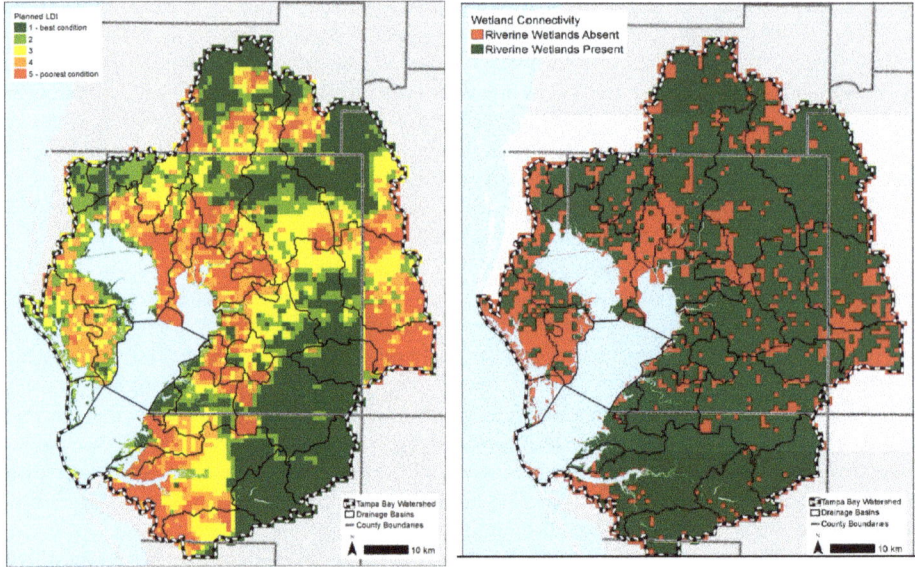

Figure 4. Example maps, projected landscape development intensity, and wetland connectivity, Tampa Bay watershed. (**a**) Planned LDI, representing economic vulnerability due to pending development (green = least vulnerable, red = most), (**b**) Wetland connectivity, showing presence or absence of riverine wetlands (green = present, red = absent).

With an overall objective of setting wetland-level priorities of restoration, preservation or mitigation, several additional measures were generated, including wetland change (by type and function), conditional assessment, hydrologic connectivity to bay, and economic vulnerability, based on planned development in the area. Regionally, the information became part of the restoration targets for the update to TBEP's Comprehensive Conservation and Management Plan. Permitting agencies adopted the historical condition and the resulting overall prioritization plan into their rules; permitting decisions reference the output to achieve regional goals one wetland at a time.

With advances in the temporal and spatial resolution of remote sensing data, the data were recently updated using the Integrated Valuation of Ecosystem Services and Tradeoffs model, or InVEST, to estimate ecosystem services values [21]. TBEP project data, which included remote sensing data for elevation (USGS Digital Elevation Model), land use and land cover (LULC), rainfall (PRISM), and watershed boundaries (USGS), were used as InVEST inputs; biophysical factors linking land cover to soil and vegetation parameters were added to finalize calculations. InVEST estimated nitrogen retention in kg/year in the TBEP watershed by basin, and relative recreational opportunity values by basin. The process required less than two days' effort, versus more than a year to build the original dataset. From a budgetary perspective, this is an investment of potentially less than $1000 versus nearly $100,000. Once the model is built, InVEST estimates can be updated with minimal effort on a regular basis, using the most recent remote sensing data.

The InVEST application, while helpful in assessing relative values, does not provide an actual estimate for the economic value of ecosystem services. The nutrient regulation model provides relative nitrogen retention by basin, which can be converted to economic equivalence using published values of $140/kg [22]. The recreation model provides an estimate of the relative recreational opportunity value, with estimates by basin/hydrologic unit ranging from a low of 4 to a high of 1937, but these values are not readily convertible to monetized estimates. Additional effort would be required to convert the relative values for recreation to economic benefits, a task likely to be beyond the budget and expertise of most coastal managers. The lack of a more complete valuation demonstrates one drawback of the InVEST approach, and is consistent with current practice of quantifying coastal ecosystem services in ways that avoid economic valuation. While not addressed in the TBEP project directly, literature on benefit-relevant indicators offers a path for translating the value estimates from the project into monetized values; see, for example, Olander [23]. From a prioritization perspective, the relative value is still helpful.

The TBEP project offers an additional example of using remote sensing data, coupled and calibrated with limited field observations to produce ecosystem services valuation information that is directly applicable to coastal management actions.

4.3. Mangrove Heart Attack

Charlotte Harbor National Estuary Program (CHNEP) had large saltwater mangrove mortality areas with adjacent areas showing stress and potential expansion of the die-off to thousands of acres; management did not know the cause, but speculated that development was impacting mangrove health. Using published values for storm protection services provided by mangroves of up to $3116/ha [24], CHNEP's mangrove area potentially offers $80 million in protection from extreme events, while past studies have found that mature, healthy mangrove in the CHNEP provide $75,300/ha in annual fisheries production alone. With ~700 square km of area, partitioning the management approach to fund appropriate mitigation measures was no small feat. In attempting to tackle the issue, staff identified published methods to use Landsat 8 color bands/normalized difference vegetation index (NDVI) to identify mangrove health in remote locations. Using green and near-infrared bands of Landsat 8 data, as per Giri et al. 2015 [25], varying conditions of mangrove could be assessed.

Staff started by comparing the Green Normalized Difference Vegetation Index data for the mangrove areas from 1985 and 2015; using the NDVI Green (NDVIg) values, natural breaks found intervals of greenness beginning at 0.9554, and gradually increasing to 0.9751. Only those which showed low NDVI green (less than 0.9554), and had declined, were included in the study. The pixel size based on the available data was $\frac{1}{4}$ acre; at the scale of the NEP, this was a useful starting point for assessing condition. Site visits were carried out to collect field samples and calibrate data at 56 sites; the field work found that the NDVI correctly evaluated the condition with 75–83% accuracy, which was more or at least equivalent to the official data that would otherwise have been used for land cover assessment (official meaning that prepared by South Florida Water Management District, or SFWMD) [26]. In total, 21,850 acres of condition-classified mangrove was mapped.

Where condition was poor, red and orange coding was used to flag areas, while magenta was used to highlight areas with low and declining condition. Figure 5 shows a section of North Captiva Island after processing. The bright colors augmented rapid review of the extensive forests throughout the study area. Many of the areas were remote and difficult to access, and aerials were insufficient to assess condition reliably.

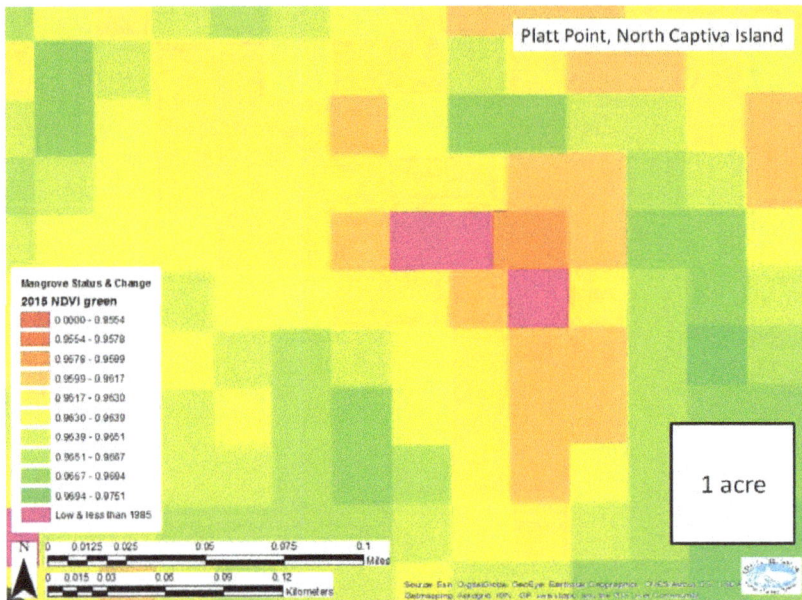

(a)

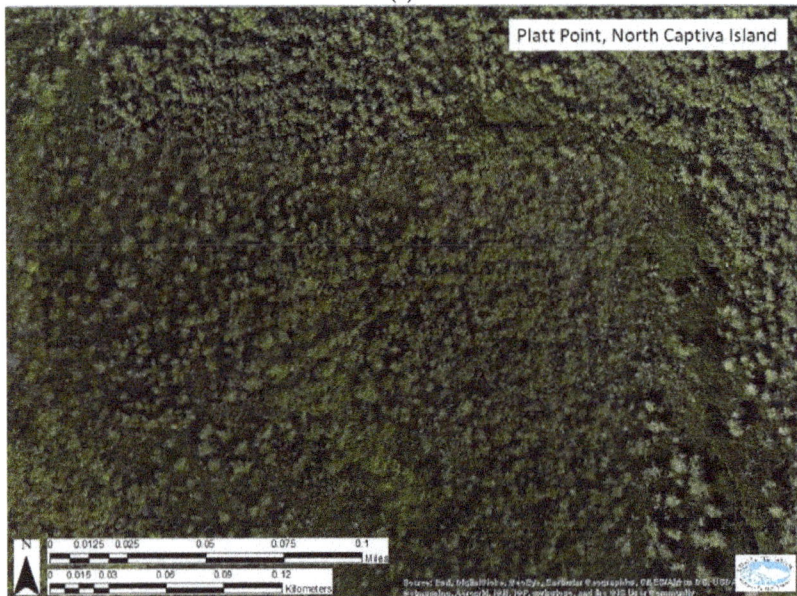

(b)

Figure 5. (a) NDVIg indicated poor condition that could not be seen from an aerial view. (b) Map: aerial view of the underlying mangrove area. Source: CHNEP [26].

At the 1:5000 scale of existing aerial photography, some decline was apparent. At 1:1000 scale, the selected pixels appeared in good condition. Upon further investigation at the 1:500 scale, which was available through Landsat 8 data, bare branches could be detected in the areas flagged as low NDVI; see Figure 6.

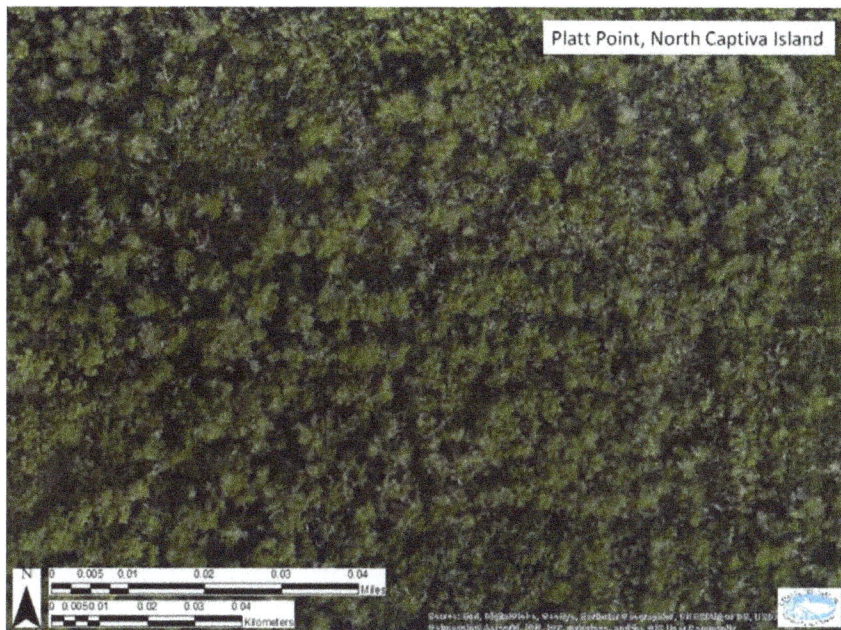

Figure 6. NDVIg indicated poor condition that could be seen at 500 scale. Source: CHNEP [26].

The authors state that "The results offer an astonishingly sensitive and detailed interpretation suggesting underlying hydrology, difficult to map from aerial photography and LiDAR digital elevation models alone . . . NDVIg was sensitive enough to identify mangrove trimming and dock construction." Two hundred sites were identified through the screening for further determination. Ultimately, staff found that natural causes (sea level rise, primarily) were generating mortality at 121 sites; 13 sites had man-made stressors with no remedy available (on private land, e.g.); 90 sites were good candidates for restoration, and 3 sites had restoration already underway. With average site size, the 103 sites with restoration targeted or underway represent potential economic savings of nearly $25 million in storm protection and $3 million annually in fisheries production. Perhaps more significantly, recent research has found that benefits of mangrove restoration may outweigh costs by 35:1 [27].

The Mangrove Heart Attack project provides a great example of using existing data to evaluate trends, perform high-level screening, and calibrate remote sensing data to site-level inspections in a spatial mapping framework. Also, because of the ongoing collection and archiving of Landsat data, no pre-restoration monitoring is required, facilitating rapid response and potentially better resource management opportunities. The CHNEP created a perfect example of using spatial mapping to identify ecosystem services linkages, and applying remote sensing data to coastal ecosystem project decisions.

5. Discussion and Conclusions

A major finding of the research is the effectiveness of case studies to illustrate the accessibility and transferability of EO data to coastal resource management needs. The case studies provided real world examples of coastal applications that calibrated remote sensing data with field observations successfully. In each highlighted case, data collection through field observations would have been prohibitively expensive and time-consuming. By methodically identifying the critical factors required for the management task at hand, coastal researchers and resource managers were able to link remote sensing data to site-scale observations and calibrate remote sensing data to confidence intervals appropriate for the application.

The outcome of the workshops was a better understanding of the requirements for building capacity in applications of Earth Observations for improved coastal management: (1) continued training opportunities in Earth Observations and models of ecosystem services, (2) emphasis on successful use-cases to demonstrate impact, and (3) improved data products specifically developed to perform well in coastal regions.

Coastal resource management frequently requires long-term monitoring, both because funders require it and to ensure methods are successful. Extrapolating or interpolating data is difficult where historical data have not been captured, and coastal environments are particularly unforgiving in this respect. The dynamics of coastal forces render data collection more expensive than on dry land, and the constantly evolving shoreline means that hindcasting is usually inappropriate. Remote sensing data may fill the gap; once calibrated, data collection pre-dating a project can be generated, with a given confidence interval.

Economic valuation of coastal ecosystems using remote sensing requires additional steps and a multi-disciplinary approach. In the case studies noted, three different approaches were used. In the first, economic values were assigned to coastal ecosystem services using the assistance of experts in non-economic disciplines to build decision rules that would provide the benefits transfer computation. In the second, InVEST models were used to provide estimates of nutrient regulation and recreation values. InVEST provides relative values, but users must assign their own estimates of economic valuation units to produce economic valuation. Finally, CHNEP summarized economic values for ecosystem services provided by mangroves, but did not attempt to quantify the change in economic values addressed by their targeted restoration efforts.

The case studies demonstrate successful applications of spatial mapping in coastal valuation environments, while the workshops reflected a broader ongoing lag in user experience and practical application to be addressed. The expanding library of successful spatial mapping applications can support future workshops, training and similar initiatives. As EO data continues to expand, there are data refinements, new algorithms, and new missions, all of which contribute to reducing some of the uncertainty expressed by the workshop participants—and support expanding use for systems monitoring, valuation and decision-making.

Remote sensing data tend to be used less when valuing coastal ecosystem services than in other ecosystems. While there are challenges to doing so, the case studies demonstrate that the gaps can be overcome. The primary obstacle perceived by coastal managers tends to relate to the geographic and temporal scale, and the case studies illustrate methods to overcome scale issues. Additional barriers include understanding the linkages between economic values and coastal ecosystem services, and how to collaborate with economists to cost-effectively bridge this gap. Finally, improved integration of and access to the plethora of remote sensing data available for use in the coastal zone may require additional training.

Author Contributions: Conceptualization, C.D.; Methodology, D.D.; Investigation, V.S.

Funding: This research was funded by NASA, Grant number 80NSSC18K0083.

Acknowledgments: The authors wish to recognize anonymous reviewers, whose comments improved the content of this article.

Conflicts of Interest: The authors declare no conflict of interest.

References

1. Barbier, E.B. A Spatial model of coastal ecosystem services. *Ecol. Econ.* **2012**, *78*, 70–79. [CrossRef]
2. Milon, J.W.; Scrogin, D. Latent preferences and valuation of wetland ecosystem restoration. *Ecol. Econ.* **2006**, *56*, 162–175. [CrossRef]
3. Whitehead, J.; Groothuis, P.; Southwick, R.; Foster-Turley, P. *Economic Values of Saginaw Bay Coastal Marshes with a Focus on Recreational Values*; Michigan Department of Environmental Quality: Kalamazoo, MI, USA, 2006.

4. Drakou, E.; Crossman, N.; Willemen, L.; Burkhard, B.; Palomo, I.; Maes, J.; Peedell, S. A visualization and data-sharing tool for ecosystem services maps: Lessons learnt, challenges, and the way forward. *Ecosyst. Serv.* **2015**, *13*, 134–140. [CrossRef]

5. Townsend, M.; Davies, K.; Hanley, N.; Hewitt, J.; Lundquist, C.; Lohrer, A. The challenge of implementing the marine ecosystem service concept. *Front. Mar. Sci.* **2018**, *5*, 1–13. [CrossRef]

6. Van der Ploeg, S.; de Groot, R.S. *The TEEB Valuation Database—A Searchable Database of 1310 Estimates of Monetary Values of Ecosystem Services*; Foundation for Sustainable Development: Wageningen, The Netherlands, 2010.

7. Drakou, E.; Pendleton, L.; Effron, M.; Ingram, J.; Teneva, L. When ecosystems and their services are not co-located: Oceans and coasts. *ICES J. Mar. Sci.* **2017**, *74*, 1531–1539. [CrossRef]

8. Van der Ploeg, S.; De Groot, D.; Wang, Y. *The TEEB Valuation Database: Overview of Structure, Data and Results*; Foundation for Sustainable Development: Wageningen, The Netherlands, 2010.

9. Barbier, E.B. Valuing Ecosystem Services for Coastal Wetland Protection and Restoration: Progress and Challenges. *Resources* **2013**, *2*, 213–230. [CrossRef]

10. Martínez-Harms, M.J.; Balvanera, P. Methods for mapping ecosystem service supply: A review. *Int. J. Biodivers. Sci. Ecosyst. Serv. Manag.* **2012**, *8*, 17–25. [CrossRef]

11. Pollack, J.B.; Yoskowitz, D.; Kim, H.C.; Montagna, P.A. Role and Value of Nitrogen Regulation Provided by Oysters (Crassostrea virginica) in the Mission-Aransas Estuary, Texas, USA. *PLoS ONE* **2013**, *8*, e65314.

12. Chung, M.G.; Kang, H.; Choi, S.U. Assessment of Coastal Ecosystem Services for Conservation Strategies in South Korea. *PLoS ONE* **2015**, *10*, e0133856. [CrossRef] [PubMed]

13. Arkema, K.K.; Verutes, G.; Bernhardt, J.R.; Clarke, C.; Rosado, S.; Canto, M.; Wood, S.A.; Ruckelshaus, M.; Rosenthal, A.; McField, M.; et al. Assessing habitat risk from human activities to inform coastal and marine spatial planning: A demonstration in Belize. *Environ. Res. Lett.* **2014**, *9*, 114016. [CrossRef]

14. Davies, K.; Fisher, K.; Dickson, M.; Thrush, S.; LeHeron, R. Improving ecosystem service frameworks to address wicked problems. *J. Ecol. Soc.* **2015**, *20*. [CrossRef]

15. Dunford, R.; Harrison, P.; Smith, A.; Dick, J.; Barton, D.N.; Martin-Lopez, B.; Kelemen, E.; Jacobs, S.; Saarikoski, H.; Turkelboom, F.; et al. Integrating methods for ecosystem service assessment: Experiences from real world situations. *Ecosyst. Serv.* **2018**, *29*, 499–514. [CrossRef]

16. Pandeya, B.; Buytaert, W.; Zulkafli, T.; Karpouzoglou, T.; Mao, F.; Hannah, D. A comparative analysis of ecosystems services valuation approaches for application at the local scale and in data scarce regions. *Ecosyst. Serv.* **2016**, *22*, 250–259. [CrossRef]

17. Lutz, P.; Musick, J. (Eds.) *The Biology of Sea Turtles*; CRC Press: Boca Raton, FL, USA, 2003.

18. Milon, J.; Scrogin, D.; Weishampel, J. *A Consistent Framework for Valuation of Wetland Ecosystem Services Using Discrete Choice Methods*; EPA Grant Number: R831598; National Center for Environmental Economics, US Environmental Protection Agency: Washington, DC, USA, 2009.

19. United States Geological Survey, or USGS. US Geological Survey. 2008. Available online: StagedProducts/ Hydrography/NHDPlus/HU4/HighResolution/GDB/ (accessed on 31 May 2019).

20. Brown, M.T.; Vivas, M.B. Landscape Development Intensity Index. *Environ. Monit. Assess.* **2005**, *101*, 289–309. [CrossRef] [PubMed]

21. Sharp, R.; Tallis, H.T.; Ricketts, T.; Guerry, A.D.; Wood, S.A.; Chaplin-Kramer, R.; Nelson, E.; Ennaanay, D.; Wolny, S.; Olwero, N.; et al. *InVEST 3.6.0 User's Guide*; The Natural Capital Project; Stanford University: Stanford, CA, USA; University of Minnesota, The Nature Conservancy, and World Wildlife Fund: Minneapolis, MN, USA, 2018.

22. Dourte, D. *Water Quality Project Cost-effectiveness: Evaluation Metrics*; Southwest Florida Water Management District: Brooksville, FL, USA, 2017.

23. Olander, L.P.; Johnston, R.J.; Tallis, H.; Kagan, J.; Maguire, L.A.; Polasky, S.; Urban, D.; Boyd, J.; Wainger, L.; Palmer, M.; et al. Benefit relevant indicators: Ecosystem services measures that link ecological and social outcomes. *Ecol. Indic.* **2018**, *85*, 1262–1272. [CrossRef]

24. Himes-Cornell, A.; Grose, S.O.; Pendleton, L. Mangrove Ecosystem Service Values and Methodological Approaches to Valuation: Where Do We Stand? *Front. Mar. Sci.* **2018**, *5*, 376. [CrossRef]

25. Giri, C.; Long, J.; Tieszen, L. Mapping and monitoring Louisiana's mangroves in the aftermath of the 2010 Gulf of Mexico oil spill. *J. Coast. Res.* **2011**, *27*, 1059–1064. [CrossRef]

26. Beever, L.; Beever, J.; Lewis, R.; Flynn, L.; Tattar, T.; Donley, E.; Neafsey, E. *Identifying and Diagnosing Locations of Ongoing and Future Saltwater Wetland Loss: Mangrove Heart Attack*; Charlotte Harbor National Estuary Program: Punta Gorda, FL, USA, 2016.

27. Groot, R.; Blignaut, J.; Ploeg, S. Benefits of Investing in Ecosystem Restoration. *Conserv. Biol.* **2013**, *27*, 1286–1293. [CrossRef] [PubMed]

MDPI

St. Alban-Anlage 66

4052 Basel

Switzerland

Tel. +41 61 683 77 34

Fax +41 61 302 89 18

www.mdpi.com

Water Editorial Office

E-mail: water@mdpi.com

www.mdpi.com/journal/water

www.ingramcontent.com/pod-product-compliance
Lightning Source LLC
Chambersburg PA
CBHW051915210326
41597CB00033B/6155